INTRODUCTION TO SEARCH ENGINE MARKETING AND ADWORDS

A GUIDE FOR ABSOLUTE BEGINNERS

Todd Kelsey

Apress®

Introduction to Search Engine Marketing and AdWords: A Guide for Absolute Beginners

Todd Kelsey
Wheaton, Illinois, USA

ISBN-13 (pbk): 978-1-4842-2847-0   ISBN-13 (electronic): 978-1-4842-2848-7
DOI 10.1007/978-1-4842-2848-7

Library of Congress Control Number: 2017945372

Managing Director: Welmoed Spahr
Editorial Director: Todd Green
Acquisitions Editor: Susan McDermott
Development Editor: Laura Berendson
Technical Reviewer: Brandon Lyon
Coordinating Editor: Rita Fernando
Copy Editor: Kezia Endsley
Cover: eStudio Calamar

Distributed to the book trade worldwide by Springer Science+Business Media New York, 233 Spring Street, 6th Floor, New York, NY 10013. Phone 1-800-SPRINGER, fax (201) 348-4505, e-mail orders-ny@springer-sbm.com, or visit www.springeronline.com. Apress Media, LLC is a California LLC and the sole member (owner) is Springer Science + Business Media Finance Inc (SSBM Finance Inc). SSBM Finance Inc is a **Delaware** corporation.

For information on translations, please e-mail rights@apress.com, or visit http://www.apress.com/rights-permissions.

Apress titles may be purchased in bulk for academic, corporate, or promotional use. eBook versions and licenses are also available for most titles. For more information, reference our Print and eBook Bulk Sales web page at http://www.apress.com/bulk-sales.

Any source code or other supplementary material referenced by the author in this book is available to readers on GitHub via the book's product page, located at www.apress.com/9781484228470. For more detailed information, please visit http://www.apress.com/source-code.

Printed on acid-free paper

Apress Business: The Unbiased Source of Business Information

Apress business books provide essential information and practical advice, each written for practitioners by recognized experts. Busy managers and professionals in all areas of the business world—and at all levels of technical sophistication—look to our books for the actionable ideas and tools they need to solve problems, update and enhance their professional skills, make their work lives easier, and capitalize on opportunity.

Whatever the topic on the business spectrum—entrepreneurship, finance, sales, marketing, management, regulation, information technology, among others—Apress has been praised for providing the objective information and unbiased advice you need to excel in your daily work life. Our authors have no axes to grind; they understand they have one job only—to deliver up-to-date, accurate information simply, concisely, and with deep insight that addresses the real needs of our readers.

It is increasingly hard to find information—whether in the news media, on the Internet, and now all too often in books—that is even-handed and has your best interests at heart. We therefore hope that you enjoy this book, which has been carefully crafted to meet our standards of quality and unbiased coverage.

We are always interested in your feedback or ideas for new titles. Perhaps you'd even like to write a book yourself. Whatever the case, reach out to us at editorial@apress.com and an editor will respond swiftly. Incidentally, at the back of this book, you will find a list of useful related titles. Please visit us at www.apress.com to sign up for newsletters and discounts on future purchases.

—*The Apress Business Team*

Contents

About the Author

Todd Kelsey, PhD, is an author and educator whose publishing credits include several books for helping people learn more about technology. He has appeared on television as a featured expert and has worked with a wide variety of corporations and non-profit organizations. He is currently an Assistant Professor of Marketing at Benedictine University in Lisle, IL (www.ben.edu).

Here's a picture of one of the things I like to do when I'm not doing digital marketing—grow sunflowers! (And measure them. Now there's some analytics for you!)

I've worked professionally in digital marketing for some time now, and I've also authored books on related topics. You're welcome to look me up on LinkedIn, and you're also welcome to invite me to connect: http://linkedin.com/in/tekelsey.

About the Technical Reviewer

Brandon Lyon is an expert in SEO, SEM, and Social Media and Web analytics, as well as President of Eagle Digital Marketing (https://www.eagledigitalmarketing.com), a full-service agency in the Chicago area. When he isn't advising local business owners and CEOs of mid-sized companies, he enjoys hockey and doing his best to survive the occasional subzero temperatures. Brandon enjoys helping companies face the challenges of the future with optimism, including navigating the treacherous waters of the Amazon e-commerce river, and taking advantage of the goldmine in marketing automation.

Introduction

Welcome to search engine marketing!

The purpose of this book is to provide a simple, focused introduction to search engine marketing, for interns who may be working at a company or non-profit organization, for students at a university, or for self-paced learners. The approach is the same one I've taken in most of the books I've written, which is conversational and friendly, with an attempt to make things fun.

The experiment is to help my readers get started with digital marketing in a way that is fun and helps you strengthen your career at your current employer, or find new work—through an internship, paid work, volunteer work, free-lance work, or any other type. The focus is on skills and approaches that will be immediately useful to a business or non-profit organization. I'm not going to try to cover everything—just the things that I think are the most helpful.

The other goal is to help you leave your intimidation in the dust. I used to be intimidated by marketing and now look at me—I'm a marketing strategist and an assistant professor of marketing! But I remember the feeling of intimidation, so part of my approach is to try to encourage readers who may feel uncertain about the field.

LinkedIn shows digital/online marketing as a top skill to have year after year, and search engine marketing is one of the core skills for digital marketing. It makes Google and other companies tens of billions of dollars a year.

Each year, the way they refer to digital marketing seems to change, but since 2013, digital marketing has been at the top. Demand will fluctuate over time, but these are skills that continuously get people hired, as you can see:

- 2014: https://blog.linkedin.com/2014/12/17/the-25-hottest-skills-that-got-people-hired-in-2014

- 2015: https://blog.linkedin.com/2016/01/12/the-25-skills-that-can-get-you-hired-in-2016

- 2016: https://blog.linkedin.com/2016/10/20/top-skills-2016-week-of-learning-linkedin

One of the things I've learned throughout my career, which I try to reinforce in these books and in my classes, is the way that the core areas of digital marketing are related. For example, search engine marketing is tightly connected to all other areas of digital marketing. The central goal of digital marketing is to develop content that flows through various channels, including through search engines.

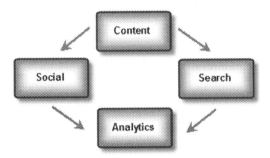

Just like each of the other areas, search engine marketing is crucial. If you do all the other steps, but you don't advertise, you won't get as much traffic, and it will be that much harder to succeed. On the other hand, as you'll learn, search engine marketing allows you to track return on investment, unlike almost every other kind of traditional advertising (radio, TV, billboards, etc.), and that's why it is so important.

You can start out small, look at how much revenue resulted from a few ads, and build your plan with confidence. Search engine marketing has helped businesses around the world, large and small, effectively sell on the Internet.

This book mentions what I call the core areas of digital marketing: Content, AdWords, Social, and Analytics (CASA). My goal is to reinforce how all the areas are connected. AdWords is Google's tool for creating ads for search engine marketing. The inspiration came from my professional background, as well as looking at the trends in the marketplace.

The Core Areas of Digital Marketing

C Content/SEO: search engine optimization is the process of attempting to boost your rank on Google so that you get higher up in search rankings when people type in particular keywords. Higher in search rankings = more clicks. The top way to boost rank is to add quality content that is relevant for your audience.

A Adwords: the process of creating and managing ads on Google (Adwords), where you attempt to get people to click on your ads when they type particular keywords in Google. You pay when someone clicks.

S Social Media Marketing: the process of creating and managing a presence on social media, including making posts, as well as creating advertisements. The main platforms are Facebook, Twitter, and YouTube, as well as Instagram and Pinterest

A Analytics (Web visitors): You can gain valuable insights when you measure the performance of your websites and advertising campaigns. Google Analytics allows you to see how many people visit your site, where they come from and what they do.

Best wishes in learning all about search engine marketing!

Introduction

This chapter takes a whirlwind tour of AdWords, a behind-the-scenes look at a real-world use of AdWords, and then we'll look at creating a Google AdWords account.

What Is AdWords?

AdWords, in a nutshell, is why Google's stock is worth so much. It's not just AdWords, it's the search engine, it's a lot of things. AdWords is part of the puzzle, and it's a tool that you use to create ads.

The magic of Google is that they figured out how to "monetize" search, which means they figured out how to make an effective online advertising program. Part of the core concept is that you can track the ad results. Unlike most other forms of advertising, with AdWords, you can determine exactly how effective your ad is. If you spent $5.00 and make $10, maybe that's a good thing. If you spent $10 and make $5, that's bad.

It's a vast oversimplification, but it's all about tracking ROI, which is return on investment. If you just want the nutshell: AdWords works. Period.

© Todd Kelsey 2017
T. Kelsey, *Introduction to Search Engine Marketing and AdWords*,
DOI 10.1007/978-1-4842-2848-7_1

Here you can see a Google search page. Go to `google.com` and search for something like tennis rackets:

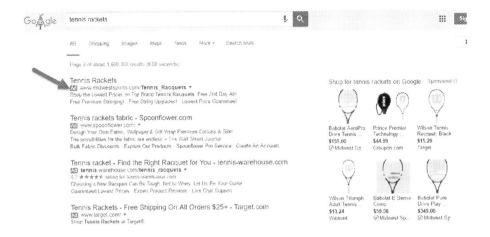

Various kinds of ads will appear, including ads with text only, and possibly ads with pictures. The ads indicate that some advertiser created them, and Google displayed them, thinking that they might be relevant to your search. Whenever you click on a link on the right, some advertiser is paying something for it, even if it is a small amount. The whole idea is, you have something to sell, you measure the clicks, and you hopefully sell what you want. Then you can continue to advertise.

Behind-the-Scenes Tour of Google AdWords

In this brief visual tour, I'm going behind the scenes at how ads are created and tweaked to display on Google. Google's AdWords tool allows you to create ads like this:

Need NASW CEU's?
Adopting the Older Child
Web ecourse. 2 continuing ed units
adoptionlearningpartners.org

Note   This is a "traditional" ad in Google. The one thing you can count on in digital marketing is that things change from time to time, and in mid-2016, Google decided to expand the length of ads, so you'll start seeing more ads like this:

GoFundMe Official Site
www.gofundme.com/ ▾
Fund Your Project On The World's #1 Personal Fundraising Website
Free To Sign Up · 5-Minute Email Support · No Goal Requirements · Mobile Friendly
Services: Crowdfunding Service, Fundraising Service, Medical Fundraising, Emergency Fundraising
How it Works · Success Stories · Sign Up Free · Education

Note that this ad is longer than the previous one. It's one in the "newer" format. But for the purpose of this chapter we're going to look at the traditional format. Don't worry about the length, just sit back and try to get a sense of how it works.

Let's go back to the original ad.

Need NASW CEU's?
Adopting the Older Child
Web ecourse. 2 continuing ed units
adoptionlearningpartners.org

The purpose is to try to get people to "click through" when they are searching on Google or looking at Gmail. Here's what each of those lines mean in the ad:

- Top line: This is the headline; it has a 25-character limit.

- Middle lines: This is the description; they have a 35-character limit.

- Bottom line: This is the display URL and it has a limit of 35 characters, but it's not clickable. It is often used to display the general address of the web site.

There's also a "hidden" line, which is known as the destination URL. It is a link, and can be as long as you like. In this case, it leads to a course page (and is 68 characters):

```
http://www.adoptionlearningpartners.org/adopting_older_child_
ceu.cfm
```

Hiding the destination URL helps the ad look cleaner.

When I create an ad, I start by gathering text in Notepad, including the title of the course and some lines that could become the title, description, and display URL. I also build the link to the course page itself.

```
--
Ethical Considerations for Social Workers: Infant Adoption: CEU Course

Need NASW CEU's?

Web ecourse. 1.0 continuing ed units.
adoptionlearningpartners.org

http://www.adoptionlearningpartners.org/ethics_infant_ceu.cfm
--
```

The title of the course was too long, so I started shortening it:

```
Ethical Considerations for Social Workers
```

Here is a small excerpt of the AdWords screen. There are a lot of options, but this is where it all comes together. You see, for example, that the course title was too long to fit in "Description Line 1":

Google shows you a little indicator bar as you're typing to give you a sense of how much space you have left:

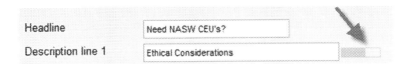

I tried a variation next. This is part of what ends up happening with Google ads. Because of the limits on how many letters/characters you can have, you have to get creative. This could involve using the two lines of description creatively, or just rewording or restating things. For example, instead of using "ethical considerations," you could write "ethics":

Description line 1   Ethics For Social Workers: Adoption

Then Google builds a preview on the right:

Once you are happy with your ad, the next step is to add keywords. These are guesses of what people might search for when they are looking for information on Google. The whole strategy of Google ads is based on trying to figure out the best combination of words that people might type.

I tried a few different options:

Basically, you pay for your advertisement based on how many people click on it. This is called "cost per click" (CPC) advertising. A thousand people might see the ad, if they type in the right keywords, but only a handful will actually click on the advertisement.

Google originally had fixed prices for keywords, but then someone had the bright idea for allowing bidding on keywords, kind of like eBay. For example, car dealers might compete and bid $5.00 or $10.00 for the keyword "Ford". If you are the highest bidder, you get the best spot for your ad.

When you're writing an ad, Google can guess how much "traffic" your ad might get:

Above, we see a CPC of $1.00. This means I'm bidding a maximum of $1.00, and I'm limiting the budget to $10.00 a day.

Google then calculates your cost using the following reasoning: Assume, for example, the cost you pay for each click will average between 44 cents and $1.00. Google assumes that you'll get between 7-12 clicks a day, so it will cost you between $5-10.

In general, what people try to do is track the results of their advertisement very carefully, so that if they spend $100 on an advertisement, they see how many widgets or books or stereos or beanie babies they sell. So if I spend $100 on my ads, I hope to sell $1000 worth of merchandise (or whatever my goal is). The goal is to get people to come to your site and buy. The strength of Google ads, unlike other forms of advertisement, is that you can track *exactly* what you get out of it. You know exactly how well the ad performs.

You can later increase your budget. You spend more on advertising and generate more revenue for your company (and help Google's stock price go higher).

If you want to change an ad, you can edit it:

Here's what a finished ad looks like. Google makes billions every year (and helps businesses around the world make billions of dollars, yen, and yuan), and this is what most of it comes down to. To click or not to click.

> Need NASW CEU's?
> Adopting the Older Child
> Web ecourse. 2 continuing ed units
> adoptionlearningpartners.org

Here's a variation of one ad, where the text in Description line 1 "wraps" down to the second line. If you can't fit the text in one line, sometimes you can stretch it to another:

Because this was a long course title, I tried wrapping it to two lines:

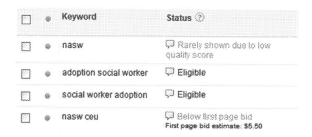

After you create your ad and choose keywords, you start to get some automated feedback from the Google system. In the first line, we see how the keyword "nasw" (National Association of Social Workers), is "rarely shown due to low quality score".

		Keyword	Status
☐	⚙	nasw	▭ Rarely shown due to low quality score
☐	⚙	adoption social worker	▭ Eligible
☐	⚙	social worker adoption	▭ Eligible
☐	⚙	nasw ceu	▭ Below first page bid First page bid estimate: $5.50

This means that Google has automatically analyzed where the ad is sending people, and if the content on the web page does not strongly match the keyword, it will get a lower quality score. The ad will therefore not show up as often. Conversely, when the keyword is more related to the page you're sending people to, you get a higher quality score. This is basically Google helping people have the best search experience.

If someone types popcorn, and you've chosen it as a keyword, but the page you send people to has nothing to do with popcorn, the ad will not display as much, because it's not relevant.

In the previous example, for the keyword "nasw," my goal is to get people who are a part of that association to take an adoption learning partners course. I have a few options for dealing with this low quality score:

- I could increase the number of times "nasw" appears on the course's web page.

- I could delete the keyword and add others. This is known as keyword optimization.

Another tidbit that Google reports to you is where your ad appears in the search results. In this example, we can see the message stating that for the keyword "nasw ceu," the bid is below what is required to display on the first page. It also reports that the first page bid estimate is currently $5.50.

Since these ads are being run with a grant from Google (Google gives free advertising to non-profit organizations under certain conditions), Google places a limit on the maximum bid a person can make with this free advertising ($1.00). So the maximum I can bid is $1.00, and for this keyword, it won't show up on the first page. Google would rather have other organizations and companies pay $5.50 each time someone clicks on their ads.

At first glance, you might think it is crazy to pay so much money when someone clicks. There's no guarantee that if someone clicks on an ad, they will buy anything.

The Bidding Game

The reason that this all ends up working is because people keep very close track of the statistics about how many people click and how many widgets they sell as a result. There's a constant "bidding game" going on between companies. They develop a web site designed to convince people to buy their product and they are able to determine how much money they can spend to sell their products and still make a profit. The game generally involves trying to get the best position for your ad, as well as to get the most conversions. *Conversions* are when people click on your ad and also end up buying your product. This is the world of online advertising.

The Guessing Game

In other forms of advertising, it's more of a guessing game. A million people drive by a billboard or see a television ad, and it's hard to say what effect can be traced directly to a specific advertisement. In traditional advertising, it's more about trying to get as much exposure as possible and hope for the best. Occasionally, you can run special promotions, like display a coupon code, so that if people use the coupon code, you know that they saw the ad and you know that it's working. But it's still not as exact. Online advertising is significantly more trackable, and therefore in some cases it is much more effective, or at least you can track the performance better. Advertisers have more confidence when they know exactly how things are performing.

We next see the search results of our keyword "nasw ceu" again. This is the second page of the search results. If you remember, Google told us our keyword was "below the first page estimate," so it would not show, or is not likely to show, on the first page. But it can show on a subsequent page.

The question is, how many people will look at the second page of results? Not as many as the first (where the bid price is higher), but some make it to the second page, and so on.

The left arrow shows we're on the second page of results; the right arrow shows our ad. It made it on the second page. Woo-hoo!

After creating a few ads, I decided to see if I could come up with a keyword that would get on the first page of results, but is still within the maximum I can spend of $1.00 per click.

I tried a variation of a phrase "nasw adoption" and the results are shown below. In the upper left, you see what it would be like if someone typed in the phrase I'm hoping they'll type in. Someone who needs to take continuing education credits—maybe they are social workers who deal with adoption.

It turns out that this keyword does result in the advertisement showing on the first page, which is a good position for the ad.

It also turns out that the "organic" search results highlight Adoption Learning Partners. Organic search results are unpaid, "natural" search results. This is where, without requiring you to spend any money, Google is out there crawling the web, categorizing it, and making automated listings. Even if you don't spend any money at all, your web site will show up somewhere in Google search results.

Notice that for this phrase, in the upper right of the screen, there are approximately 80,000 links that Google will be glad to show you. In this case, the Adoption Learning Partners site has a pretty good ranking, because of the number of people who visit the site, and because of how relevant the web page is—it has courses that are "NASW approved".

It's not necessarily easy to get the top "unpaid" results, especially when there's a lot of web sites that have similar information or when web sites are in competition with each other. SEO (search engine optimization) is the process of getting your web site to show up in a better spot on Google without spending any money.

It's great when you can get people to come to your site without clicking on an ad that's going to cost you money. But it's harder to do this, and it's not always as reliable as paid advertising.

Paid advertising is known as *search engine marketing*. There are different philosophies—some companies swear up and down that you should never pay Google a dime, that their company can get you the best ranking, or that if you buy their secret guide, it will reveal all the latest secrets. Then there are those who are loyal to SEM, where you pay for the advertisement, and don't even bother trying to do SEO.

In general, it's good to do a bit of both. In the end, paid advertising is more solid than search engine optimization. Google exists to make money, so there are ways that they encourage businesses to use the paid advertising. They might not give your business a high rank—they offer no guarantees to anyone. But they'd naturally prefer you to give them your money. It pays to consider that Google is a business, so SEM is generally a solid way to go.

Important Note Things change!

I try to say this as often as I can, in as many ways as I can. Google is especially active in constantly improving their products. There's a fair chance that by the time you read this book, there will be some changes and updates to Google AdWords, in strategies, and so on. The goal of this book is to introduce you to AdWords, not to include every exact feature and change to Google's products.

For example, there is a traditional ad format for Google ads, which this book is based on, which looks something like this:

Need NASW CEU's?
Adopting the Older Child
Web ecourse. 2 continuing ed units
adoptionlearningpartners.org

If you go into any AdWords account, chances are you will encounter ads using this format. But they are also experimenting with a new format. Will it stay? Perhaps.

It allows advertisers to have more text in the ad.

Tennis racket - Find the Right Racquet for You - tennis-warehouse.com
[Ad] www.tennis-warehouse.com/tennis_racquets ▾
Choosing a New **Racquet** Can Be Tough. Not to Worry, Let Us Be Your Guide.
Categories: Equipment & Accessories, Shoes, Apparel, Gift Cards
Accessories: Sensors And Gadgets, Tennis Balls, Tennis Towels, Sunglasses, Court Equipment

But you don't have to use all the options, and you may see an ad like this:

Tennis Rackets - Everything for Tennis Lovers - tennisexpress.com
[Ad] www.tennisexpress.com/ ▾
Tennis Rackets. Huge Selection. Lowest Prices. Free Shipping and Free Returns*!

Don't worry about format yet. Just launch in and try things and be aware that *things change.* Hopefully for the better.

Creating an Account

If you don't have an AdWords account already, go and create one! Go to `http://www.google.com/adwords` to get started.

After you've created an account, it may ask you to create an ad as part of the "wizard" process. That's okay, just be sure to pause the campaign after you are into AdWords.

Note   Make sure to pause your campaign once you start your AdWords account!

Look for the little green button next to your campaign. If you don't pause it, you could end up spending money you don't have! Click on the green button and switch to pause.

Click on the green dot and select Paused:

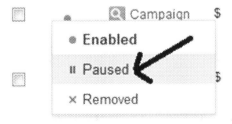

Getting Help

The Google AdWords Help Center is extensive and worth checking out. See https://support.google.com/adwords. I recommend exploring all the links at the top (the Community Forum and the Contact Us) to get a sense of contact information.

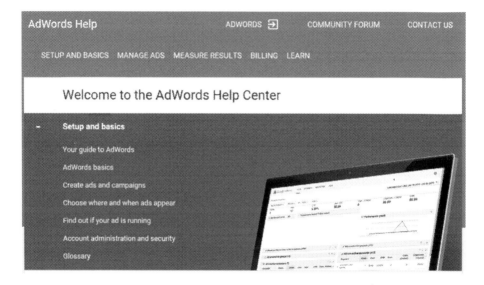

You can also scroll down and click on the +/- links to see more topics.

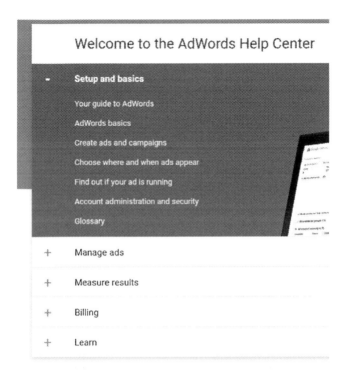

Conclusion

Dear Reader,

Thanks for reading this chapter! Don't be alarmed if you don't feel like you "get it" yet—we're just taking a look at things, getting to know how things work. In subsequent chapters, you'll be getting to know AdWords better. Before you know it, you'll be creating ads and getting clicks!

Cheers,

—Todd

Quickstart

The purpose of this chapter is to give you a chance to get right into things. It assumes that you have an AdWords account, and that the billing is set up.

Creating a Campaign

So, again, this chapter assumes you've read the first one, that you've started an account, and that billing, etc. is ready to go.

So sign in to AdWords (`www.google.com/adwords`) and click on the Campaigns tab:

Campaigns

Depending on the account (Google routinely changes the way things work), you might have a Create Your First Campaign button, or you might just need to click on the standard Campaign button under the Campaign tab.

© Todd Kelsey 2017
T. Kelsey, *Introduction to Search Engine Marketing and AdWords*,
DOI 10.1007/978-1-4842-2848-7_2

There are a variety of types of campaigns, but for now, just keep it simple and choose Search Network Only.

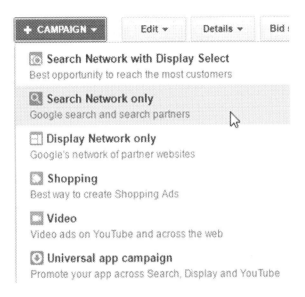

And at this juncture, one thing I want to mention is my own experience.

I used to be intimidated by Google AdWords. I'd heard about search engine marketing, had talked to people who had done it, and had also heard a bit about search engine optimization. But it seemed like a mysterious world that only elite people could deal with.

The funny thing is, Facebook Advertising helped me get into Google AdWords. For whatever reason, I wanted to advertise this thing called the Sunflower Club, so I looked at Facebook advertising, and it was so darn simple that it helped me build confidence. Although I'm firmly convinced that Google rocks, there are other entities in the world of online advertising. Personally I hope Google buys Facebook or ends up helping them with advertising. Anyway, I tried Facebook advertising, which helped me get my feet wet. Check out this introduction to it—http://tinyurl.com/fbadv. (Notice that I used a tiny URL? Guess what the link leads to? You guessed it, a Google doc.)

The point of these tutorials is to make things user-friendly. Don't be intimidated. If I can do it, you can do it.

Google provides a "wizard" to guide you through the process:

- *Select campaign settings*: Choose the main general settings, such as budget, how much you are willing to pay for a click, and so on.

- *Create ad groups*: Google has a hierarchy for organizing ads and campaigns. Just think of a campaign as the biggest box, and then the medium size box is an ad group, and in each ad group, you can place ads. Practically speaking, the ad group allows you to try multiple ads in it, which would have some similarities.

- *Create ads*: This is where you create the ad itself.

- *Review ad groups*: In this wizard format, Google gives you a chance to review things before you set the campaign in motion.

Sometimes I use the wizard and sometimes I "exit" the wizard, but it might help to explore it. In many cases, Google provides "default choices" to make the process easier and nine times out of ten, you can just go with the default settings until you learn more.

In the campaign Type section, click on the All Features radio button.

Then, give your campaign a name in the Campaign Name field. Any name will do. You can even use Sunflower Club if you like. Don't worry about all the options right now.

Then, just start scrolling down.

Locations and Languages are cool options. You can scroll by most of this stuff, but you can always change it if you want.

Networks ?   To choose different networks, edit the campaign type above, or create a new campaign.

       ✓ **Google Search Network** ?
          ✓ Include search partners

Devices ?   Ads will show on all eligible devices by default.

Locations ?   Which locations do you want to target (or exclude) in your campaign?

       ○ All countries and territories
       ● United States and Canada
       ○ United States
       ○ Let me choose...

       Enter a location to target or exclude.   Advanced search

       For example, a country, city, region, or postal code

Google rocks partly because they're on a constant world tour. They're global friendly. If you wanted, for example, you could run an ad campaign in Chinese (and you should probably learn how to do that).

Languages ?   Choose the language of the sites that you'd like your ads to appear on. Be sure to write your ads in the language that you target, since AdWords doesn't translate ads or keywords.

COOL!

☐ All languages		
☐ Arabic	☐ Greek	☐ Portuguese
☐ Bulgarian	☐ Hebrew	☐ Romanian
☐ Catalan	☐ Hindi	☐ Russian
☐ Chinese (simplified)	☐ Hungarian	☐ Serbian
☐ Chinese (traditional)	☐ Icelandic	☐ Slovak
☐ Croatian	☐ Indonesian	☐ Slovenian
☐ Czech	☐ Italian	☐ Spanish
☐ Danish	☐ Japanese	☐ Swedish
☐ Dutch	☐ Korean	☐ Thai
✓ English	☐ Latvian	☐ Turkish
☐ Estonian	☐ Lithuanian	☐ Ukrainian
☐ Filipino	☐ Malay	☐ Urdu
☐ Finnish	☐ Norwegian	☐ Vietnamese
☐ French	☐ Persian	
☐ German	☐ Polish	

Remember, if you like, click on whatever you want. You can't hurt anything and it will help you learn more about Google's options.

Bid and Budget (Something Really Important, Part I)

This section is really important.

I'm advocating you try AdWords, just to put something out there and get a taste for it. Don't be too concerned about the web page itself, or how perfect the ad is, etc. I recommend just trying it so you can say you did. If you're anything like me, you'll think, okay, so that wasn't so bad! Then you can build from there.

Note One very important thing when trying it out, is to *be aware of the budget!* In the Bidding and Budget section, be sure to set the budget to a low amount.

I suggest you enter $5 a day into the Bidding and Budget section. That means, in order to learn how to use AdWords, you might end up with some free advertising budget (they have all kinds of offers they'll e-mail you about), or you you might spend $25-50 in the end. This isn't a bad deal to learn to use AdWords.

Don't worry about manual bidding or automatic bidding, and don't be frightened. You might see a screen like this:

Bid strategy ? Choose how you'd like to set bids for your ads.

Manual: Manual CPC ▾

You set your own maximum cost-per-click (CPC) for your ads.

Enable Enhanced CPC ?

Unavailable because conversion tracking isn't set up. Learn more

Default bid ? $

This bid applies to the first ad group in this campaign, which you'll create in the next step.

You can choose manual bidding. Use $1.00 as a default bid to keep it simple. Click on all the little question marks in AdWords to keep learning and keep rolling.

AdWords Is Like eBay

AdWords is kind of like eBay. It's an auction situation, where you are *bidding* on how much you're willing to pay to get someone to *click*. Bidding for clicks. You create an ad and you try to get someone to click on it. (Go to Google, type in tennis rackets, and check out the ads. They're all trying to get you to click on them, and they are all going to pay something to Google if you do.) Keep in mind, when you create a budget, to set a bid price for your click. That doesn't mean Google is just going to take $5 from you—you actually have to get people to click on an ad, and that's the main part of the AdWords story.

You can click on that drop-down menu and decide how to do business. When you start a campaign, when you go into the settings, Google might automatically set the bidding to automatic. That's fine. That's good for Google because it probably means they will set your bid higher—high enough to get clicks, which burns up more money.

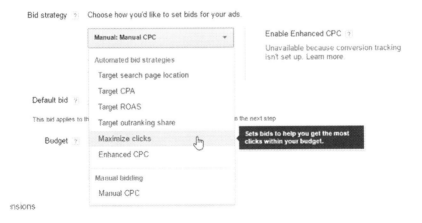

If you like, set it to Maximize Clicks and see what happens. This is also a way to keep it simple at first.

AdWords Is Doable

It's doable. Just remember that the great thing about AdWords is that it allows you to:

> *Track effectiveness*
>
> which is also known as
>
> tracking ROI
>
> and ROI = return on investment
>
> That's magic for a business
>
> It keeps a business going and makes it money
>
> That keeps you employed

As you scroll on down the Campaign screen, one of the final stops is the Ad Extensions area:

Ad extensions

You can use this optional feature to include relevant business information with your ads. Take a tour

Location	?		Extend my ads with location information
Sitelinks	?		Extend my ads with links to sections of my site
Call	?		Extend my ads with a phone number
Callouts	?		Extend my ads with additional descriptive text
Structured snippets	?		Extend my ads with structured snippets

Save and continue   Cancel new campaign

Pretty cool stuff. Ad Extensions are ways of expanding ads on Google; for now you can ignore this section. Keep it simple. But the Advanced Settings contain another really important thing: the schedule.

Advanced settings

⊕ Schedule: Start date, end date, ad scheduling
⊞ Ad delivery: Ad rotation, frequency capping
⊞ Demographic bidding

Ad Campaign Schedule (Something Really Important, Part Two)

Okay, this is also really important. You'll always want to be aware of the schedule of your ad campaign.

For example, if you set your ad campaign to $5/day and connect it to your credit card, but you don't set an end date, Google will gladly burn through your $5/day, and if you lose interest in online advertising and ignore your bank statement and then find a year later that you've burned up $1,500 in advertising. Well, that would not be fun.

Check out the advanced settings area. Part I of important things was the amount of daily budget. Part II is the schedule. Set an end date. I suggest one or two days to start. The cost of a movie.

In advanced settings, to the left of the Schedule Start Date Link, there's a + or – sign. Click on it. It expands and collapses the section.

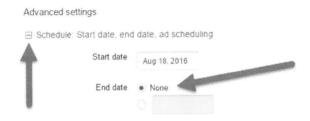

Notice that Google sets the end date to None. They assume that you will want to have an endless relationship. Maybe so. But for now, set an end date. (You can also click on the Start date if you want to change it.)

Click the blank radio button, the second one down.

Then roll your mouse over the blank field and click it:

Choose a date—one or two days from the day you start the campaign is a good suggestion.

Then click Save and Continue:

Save and continue

Now Back to AdWords

Okay, so back in AdWords we go.

As you go along in the wizard when you're creating a campaign, the next page you get to is the one where you create an ad. (Don't worry about naming the ad group; it is just a container for ads.)

You can create different kinds of ads, but for now just leave it on "text ad". Before you do anything else, I recommend clicking on all the little question marks. It's a good habit to get into as you learn to use the program.

Create an ad

● Text ad ○ Dynamic search ad ○ Mobile app engagement ○ Call-only ad

Final URL ?

Headline 1 ?

Headline 2 ?

Path ? www.example.com
/   /

Description ?

The Final URL is the link of the page you are pointing the ad at, and when you start typing in text for a Headline, it will tell you how many characters you have left.

Final URL ? www.rgbexchange.org

Headline 1 ? Come and Visit| 15

The line I originally wanted was one character too long:

Come and Visit the RGB Exchange   -1

The nice thing about AdWords is you can see an immediate preview along with sample text that appears:

Ad preview: Your ad might look different on various browsers or devices. For example, its text may be shortened to fit a narrow screen. Learn More

Mobile ad

Come and Visit the RGBExchange - A New Way to See Non-Profits
Ad www.rgbexchange.org

Come and see how we're developing a stock exchange for non-profit organizations

Desktop ad

Come and Visit the RGBExchange - A New Way to See Non-...
Ad www.rgbexchange.org
Come and see how we're developing a stock exchange for non-profit organizations

Don't worry too much about the options right now; just try writing a headline that gets attention and keep it simple. Write a simple description of what you're advertising.

Final URL [?]

Headline 1 [?]

Headline 2 [?]

Path [?] www.example.com
/   /

Description [?]

Here's a summary of what the different parts of an ad do:

- *Final URL:* This is a link to where your ad is going (such as your blog, a web site, or an individual page on a web site)

- *Headline 1 and 2:* You get to have two headlines; if you look at the ad preview, you'll see that the longer headlines display better on mobile devices. You also need something for the second headline, but you don't have to use up the entire space.

- *Path:* You can leave this blank until you are more experienced in AdWords; it allows you to display a section of a web site such as rgbexchange.org/nonprofits right in the ad, if desired.

- *Description:* This is where you provide a simple description of your value proposition. Why would someone want to visit your page? What will be there when they get there? A good way to learn about descriptions is to Google various topics (tennis rackets) and see how people write the ads.

If you're just learning and you don't have a blog or web site to advertise, you're welcome to make an advertisement for www.rgbexchange.org to practice. You can use the text above or any text you like.

Keywords

The way Google AdWords works is that you create a text ad, and then eventually you have to decide what words you want to connect to it.

If you think of a Google search, you might type in a work or phrase, such as "tennis rackets".

Then you type in that *keyword*, and any number of ads display.

Shop for tennis rackets on Google   Sponsored

Babolat AeroPro ...	Wilson Tennis Racquet,	Head MicroGel Radical Mid ...	Prince Premier Technology ...	Babolat E Sense Lite ...	Babolat E Sense Comp
$159.00	$15.29	$69.00	$44.99	$59.00	$59.00
⊘ Midwest S...	Target	Walmart	Groupon.com	⊘ Midwest S...	⊘ Midwest S...

Tennis Racquets - Choosing New Racquets Can Be Tough
Ad www.**tennis**-warehouse.com/**Racquets** ▾
4.7 ★★★★☆ rating for tennis-warehouse.com
Let Us Be Your Guide.
Expert Product Reviews · Guaranteed Lowest Prices · Live Chat Support

Tennis Rackets Ship Free - Huge Selection, Discounted Prices
Ad www.**tennisexpress**.com/**Racquets** ▾
4.6 ★★★★☆ rating for tennisexpress.com
Free Shipping and Free Returns*!
Order Today, Ships Today · Free Shipping & Returns*

If you click on one of those ads, the company who placed that ad *pays* something.

> *Advertiser:* When you make an ad, you choose keywords related to your ad that might trigger your ad. And you tell Google how much you're willing to bid to get someone to click on the ad.

> *Google User:* When you do a Google search, you type in a phrase for something you're looking for, such as tennis rackets. Ads display, and depending on how compelling the ad is to you, you choose an ad to click on.

What you do is determine this: "When people type in my keyword, I'm willing to pay this amount for a click."

There's a whole art and science around keywords, but for now, keep it simple. If you are using the RGB Exchange ad as an example, just type in a keyword phrase like "non-profit organizations" in the keyword area.

Keywords

⊟ Select keywords
Your ad can show on Google when people search for things related to your keywords.

Tips

- Start with 10-20 keywords.
- Be specific: avoid one-word keywords. Choose phrases that customers would use to search for your products and services.
- By default, keywords are broad matched to searches to help you capture a wider range of relevant traffic. Use match types to control this.
- Learn more about choosing effective keywords.

Enter one keyword per line.

```
nonprofit organizations
```

Then click Estimate Search Traffic:

> **Estimate search traffic**

Google chews on that and then gives you some information, depending on how much competition there is for that keyword. It's an auction process, in effect.

Some companies are willing to pay top dollar to get the first position, that is, the right column, on the first page of the search results. Depending on your bid, and some other factors you can learn about in other tutorials, you'll get on the first page, second page, and so on. Money talks, so the higher the bid, the better the position.

The general idea with AdWords is, you should have something to sell, keep track of how much money you're spending on ads, and see how effective it is.

Estimated traffic summary ?

The following are approximations for the keywords above.
Based on max CPC: **$2.00** and budget: **$329.00/day**.

Avg. CPC: $0.87 - $1.06

Clicks/day: 16 - 19

Cost/day: ? $15.60 - $19.07

Translation:

> CPC stands for cost per click. The bidding places the
> cost of a click between 36 cents and $1.25 in this case.
> It varies significantly for different keywords. Google
> also estimates the number of clicks a day you could
> get, and then gives a cost.

Congratulations on getting this far. Click Save Ad Group:

Save ad group

If all goes well, your ad will be created.

But don't be surprised if you get a screen like this.

 ## Advertising Policies

Please carefully review all the advertising
your ads, keywords, website, and accour
result in your AdWords account, and any

The ultimate goal of the Google AdWords
providing a great user experience is the fii
will help you meet this goal and be more

Application of our policies will always
the right to reject or approve any ads.
represent and warrant that your advei

Ad creative guidelines

Grammar and text layout

Gimmicky use of text

Competitive language

Google may ask that you Please Correct the Errors Below:

Please correct the errors below

It will tell you why.

⚠ This ad does not meet our editorial guidelines. Please correct these problems:

- **club!:** Punctuation - nonstandard punctuation. ▸ Details

If you click on details, it will explain.

○ Google policy does not permit excessive or unnecessary punctuation or symbols, or use of nonstandard punctuation, including tildes (~), asterisks (*),and vertical rules (|). Please see our full policy.

Recommended actions: ○ Please delete or edit this word.
○ You can also request an exception.
☐ Request an exception.
Your ad will be saved and reviewed by AdWords staff.

If your ad said "Join da club," for example, you might need to change it to Join the club:

Sunflowers? Join the club

Then, you click save and finish.

When all is well, you'll see something like this:

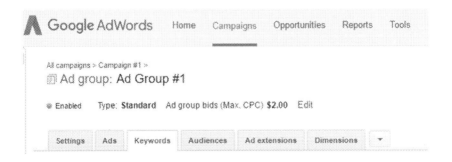

Under the Keywords tab, it will look something like the following:

Don't be alarmed! Millions of people have learned to use AdWords, and so can you. It's just a matter of trying things out, learning about keywords and ads, and choosing some of the details.

One thing that's helpful to do is to practice clicking around on the various tabs. You can click on the Ads tab, for example.

It might take some time before the ad is approved, but eventually, you'll want to come back and look at how things go.

Don't worry about this too much. After your campaign runs a day or two, come back and look and see how things are going.

Here we're looking at the start—before any data has been generated, before the ad has run, and before anyone has clicked on the ad.

Clicks ?	Impr. ?	CTR ? ↑	Avg. CPC ?	Cost ?	Avg. Pos. ?
0	0	0.00%	$0.00	$0.00	0.0

What you'll be interested to see is how many clicks there are. The Impr column is Impressions; try clicking on its question mark, as well as on the CTR question mark, to explore what they mean.

Impressions estimate how many people glanced at your ad, whereas Clicks are what you pay for.

Accessing a Campaign and Ad

If you get lost, sign in to AdWords and click on the Campaigns tab:

(If you don't see it, click on the Campaigns or All Campaigns link at the top.)

Then click on the Keywords or Ads tab:

The Ads tab will show you whether your ad is "pending review" or not. An ad may be pending review for an hour or a day; it's hard to say. Generally this process is fairly quick. It just depends.

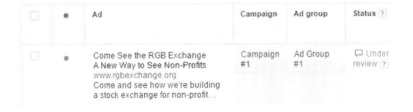

Pausing a Campaign

In the Ad area, you can click on the green dot's drop-down menu next to an ad and pause it.

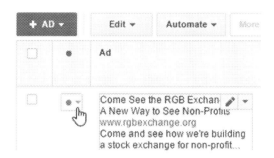

Chose the Paused option:

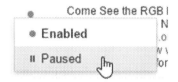

The ad will then show up as paused:

Click to enable it again. Try it!

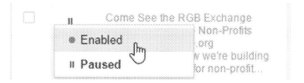

You can also pause an entire campaign.

> *Campaigns:* The biggest "container" for ads.

> *Ad groups:* Within a campaign, you can have several ad groups. Various ad groups can target different web sites, languages, people, etc.

> *Ads:* Within an ad group, it's common to create several ads, to try out various headlines and descriptions, and you can look back and see which worked best. An ad group shares keywords, so another purpose of ad groups and sets of ads within them is to focus on particular sets of keywords.

Settings

The Settings tab is the place where you can adjust options like when your campaign ends.

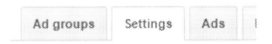

Click on Settings, then click on Advanced settings. Click + or − sign next to Schedule: Start Date to expand the area if you don't see the section below.

When you see the section, click the Edit link by the end date:

Conclusion

Dear Reader,

You made it through this quickstart section. Congratulations!

Thanks for checking this out. If you like, join the AdWords 101 LinkedIn group (www.linkedin.com).

I also recommend checking out http://www.google.com/adwords/beginners guide/en-US.

Cheers,

—Todd

Launching a Campaign

This chapter takes a closer look at planning and launching a simple ad campaign, and reviews some of the related options in AdWords.

Planning the Ad

The first step I take when planning an ad is to look at the content on the page and site. In most cases, you will have an entire site, with various kinds of content that you send people to using an ad. You might even create a new page, or a *microsite*, with a specific message as part of a campaign.

This is a simplified example, and I'm approaching it from the standpoint of thinking about what kinds of things to put in the ad itself. I'm basing that on the page content.

To plan an ad, you can go into AdWords and create a new ad. If you don't want to go to the trouble of having to create a campaign, etc., you can use a "prototyping" tool. There may be better ones out there, but I made a simple here: `http://tinyurl.com/adwords-proto`.

© Todd Kelsey 2017
T. Kelsey, *Introduction to Search Engine Marketing and AdWords*,
DOI 10.1007/978-1-4842-2848-7_3

It allows you to enter the text and it has the character lengths built-in, so if you go over the limit, you can edit and change things.

Last minute news flash: Also try this tool: http://www.blastam.com/ expanded-text-ad-preview-tool/.

As an exercise in planning ads, begin by looking at your target—the area where you are planning on "pointing" your ad, as well as the site, page, or product you are planning to advertise. I recommend using a web site or blog and considering what kind of value proposition you can share about the site. Take notes on unique things about the site, such as features, price, etc. Things people might want to know about. If you need an example, you're welcome to use www.rgbexchange.org.

Use Google to look at how other people communicate their value propositions.

Tennis racket - Find the Right Racquet for You - tennis-warehouse.com
tennis-warehouse.com/**tennis_racquets** ▾
4.7 ★★★★★ rating for tennis-warehouse.com
Choosing a New Racquet Can Be Tough. Not to Worry, Let Us Be Your Guide.
Guaranteed Lowest Prices · Live Chat Support · Expert Product Reviews

Tennis Racquets - DICKSSportingGoods.com
www.dickssportinggoods.com/Sports ▾
Shop Tennis Racquets Today. Available At DICK'S Sporting Goods!
Ratings: Shipping 9/10 - Service 9/10 - Selection 9/10 - Returns 9/10 - Product quality 9/10
Top Sports Brands · Sports Equipment · This Week's Deals · Store Locator

I recommend trying to make three different ads that point to the same site. Try the tool and techniques mentioned previously and write the text of the ads before going into AdWords.

Don't Forget Google

Google also has a nice article that can help you get started, which includes a short video. It's available at: https://support.google.com/adwords/answer/1704392?hl=en or http://tinyurl.com/writingadwords

Creating the Campaign

To create a campaign, sign in to AdWords and choose Campaign.

Choose Search Network Only.

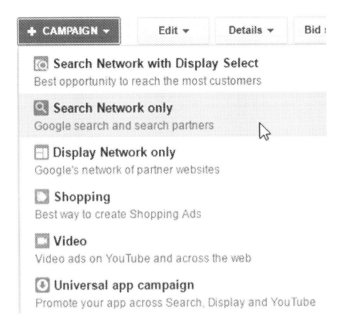

Choose All Features:

Enter a campaign name that you can easily remember:

Confirm the location you want to target. Google will guess for you, but you can always change it. Remember you can click on the little question mark.

Locations ? Which locations do you want to target (or exclude) in your campaign?
  ○ All countries and territories
  ● United States and Canada
  ○ United States
  ○ Let me choose...

  Enter a location to target or exclude   Advanced search

  For example, a country, city, region, or postal code

Bidding and budget is an important area. Remember to click on the question marks for more information!

Bid strategy ?   Choose how you'd like to set bids for your ads.

| Manual: Manual CPC | ▾ |

You set your own maximum cost-per-click (CPC) for your ads.

Enable Enhanced CPC ?

Unavailable because conversion tracking isn't set up. Learn more.

Default bid ? $ [   ]

This bid applies to the first ad group in this campaign, which you'll create in the next step.

Budget ? $ [   ] per day

Actual daily spend may vary. ?

Start by setting the bidding to Manual CPC (cost per click). If you plan to actually run a campaign, I suggest a daily budget of $5 and a default bid of $1. Remember that you can create an entire campaign with multiple ads, experiment with keywords, and then pause the campaign, if you just want to learn. At some point I recommend actually setting your campaign in motion.

If it isn't clear yet, one point to keep in mind is that when you set your budget, it's just the maximum; it doesn't necessarily mean you'll spend that much money. When you are using the CPC (cost per click) model, you'll only spend money when people actually click on your ad. Remember to be aware of your daily budget.

If you are a beginner, don't worry about ad extensions at the moment. When you have time, take the tour or click on any other link you can find such as the question marks, to find out more about ad extensions.

Ad extensions

You can use this optional feature to include relevant business information with your ads. Take a tour

Location ?	☐	Extend my ads with location information
Sitelinks ?	☐	Extend my ads with links to sections of my site
Call ?	☐	Extend my ads with a phone number
App ?	☐	Extend my ads with a link to a mobile/tablet app.
Reviews ?	☐	Extend my ads with reviews
Callouts ?	☐	Extend my ads with additional descriptive text
Structured snippets ?	☐	Extend my ads with structured snippets

Choose the Start/End Date and Schedule

Okay, time to wake up! This is a very important section.

You need to be *certain* that you click on the schedule link, because you want to make sure that you consciously choose a start/end date and a schedule.

Advanced settings

⊟ Schedule: Start date, end date, ad scheduling

If you don't, Google will be glad to gobble up your bank account on an ongoing basis. Until/unless something changes, the default setting is for there to be no end date.

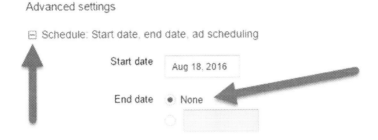

Todd: *"Naughty Google!"*

Google: *"Who, little old me?"*

To remedy this situation, just click None:

Then click in the field and choose a specific end date.

Ah, much better. This is one of the most important things to remember. (That's why I'm repeating myself!)

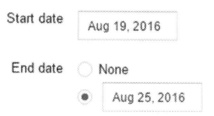

You can skip over ad delivery and demographic bidding, unless you want to explore them. When you're ready, click the Save and Continue button.

Advanced settings

⊞ Schedule: Start date, end date, ad scheduling

⊞ Ad delivery: Ad rotation

⊞ Dynamic Search Ads

⊞ Campaign URL options (advanced)

Save and continue   Cancel new campaign

Congratulations! You've created a campaign. Now you're ready to create an ad.

Creating an Ad

The first thing you do is choose a name for the ad group.

Name this ad group

An ad group contains one or more ads and a set of related keywords. For best results, try to focus all the ads and keywords in this ad group on one product or service. Learn more about how to structure your account.

Ad group name.

Ad Group #1

You'll probably want to click on the Learn More link as well.

Enter one of the ads that you created during the planning stage.

Final URL ?

Headline 1 ?

Headline 2 ?

Path ?   www.example.com

Description ?

Use this example for practice:

Final URL: www.rgbexchange.org

Headline I: Learn about 2nd Gen Donations

Headline 2: Visit the new RGB Exchange

Description: Come and see what is like to treat donations like investments, in a portfolio

Keywords

The next stage is to choose keywords; these are simply the words you think someone will type in when they search on Google, in which the product/page/message you have is *relevant* to their search.

Although it's good to try a lot of variations, it's also important to remember that the "scattergun" approach is not necessarily the most effective. You want to choose keywords that have the most direct and relevant connection; that is, if someone types them in, they're most likely to find your site.

Another key point to keep in mind is that live humans aren't the only ones doing this. Google is also checking on relevance. Google assigns a "quality score" by checking where you're sending people to and determining how relevant the content at your site is to the keywords. The more relevant the content, the better the quality score, and the better "position" your ad will have. Ad position is set by a number of factors (including your bid).

Suffice it to say, go for relevance.

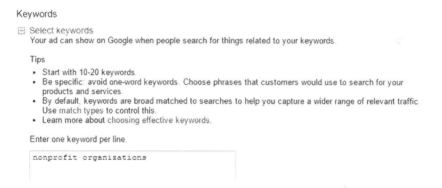

Think of keywords that relate to the site or product you are promoting. Try to come up with 5-10. If you are following the example, take a look at www. rgbexchange.org and think of keywords related to non-profits, charities, or questions people have about donations.

You can also use the AdWords feature has for *suggesting* keywords.

On the right side of the keywords area, you can click Add next to suggested keywords:

⊟ Category: Profit Listings
    « Add all from this category
    « Add   non profit organizations list
    « Add   list of non profit organizations
    « Add   non profit list
    « Add   list non profit organizations
    « Add   international non profit organizations
list
    « Add   not for profit organizations list
    « Add   non for profit organizations list
    « Add   list of not for profit organizations

You can also click on the -/+ icon in the upper left to open and close categories:

⊞ Category: Profit Listings

⊞ Category: Organisation Non

⊞ Category: Profit Start

⊞ Category: Profit Examples

⊞ Category: Not For Profit

⊞ Category: Nonprofit

⊞ Category: Nonprofit Organization

⊞ Category: Profit International

⊞ Category: Bylaws Non

⊞ Category: Profit Corporation

When you enter your final URL, Google will scan the site and come up with suggestions. Try to come up with a few on your own, and then try adding 10-20 keywords from the suggested area.

Your list might end up looking something like this. You can then click on Estimate Search Traffic:

Enter one keyword per line.

```
non-profits
list of non profit organizations
list non profit organizations
not for profit organizations list
a list of non profit organizations
not for profit organizations
what is a not for profit organization
top not for profit organizations
```

Estimate search traffic

The results depend largely on what keywords you chose and what your budget was. In this case, the budget is 329/day because it is a Google grant, where Google provides ad budget for qualifying non-profit organizations, starting at 329/day.

Re-estimate search traffic

Estimated traffic summary ?

The following are approximations for the keywords above.
Based on max CPC: **$2.00** and budget: **$329.00/day**.

Avg. CPC: $0.65 - $0.80

Clicks/day: 168 - 205

Cost/day: ? $122.14 - $149.28

The Average CPC is an attempt to estimate what the cost per click might be. If your potential cost per click is .80 cents, and your daily budget is $5.00, you can get six clicks per day (6 x .80 = $4.80).

In this section, Google looks at how much traffic there is for actual keywords. Remember, there has to be people out there searching for a keyword in order for Google to actually display your ad. Then it has to have a high enough bid to display (you just have to try), and it has to be compelling enough for people to click on. But it's doable, and millions of people click on ads every day.

When you're ready to go, click Save Ad Group.

Congratulations! You ad has been created. Most likely you will be returned to the keywords area in that ad group.

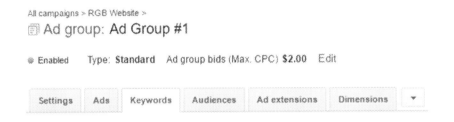

Ads and Keywords: They All Get Clicks

One of the important things to understand about AdWords is that in addition to looking at the performance of ads, you get to see how individual keywords perform. Some or all of the keywords you choose might get clicks, and that determines how many clicks the overall ad gets.

The process of optimization and monitoring performance is partly a question of looking at what worked and what didn't, so when your campaign is running, each keyword will perform differently.

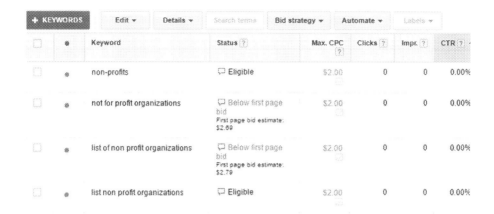

Some keywords might get more clicks than others.

Reviewing Your Keywords

Even before your ad officially launches, you can get some idea of what's going on with keywords. You might see messages about quality score. It might also alert you to the fact that at your current bid, the ad might not show on the first page. It doesn't mean people won't see it, but they won't see it on the first page of the search results.

Notice in the previous graphic that the first page bid estimate for the keyword phrase "list of non profit organizations" is $2.79.

To place these ideas in context, if we go to the Google search page and look at the paid ads on the right, we can guess that the advertisers on the first page are willing to pay as high as $2.79 per click.

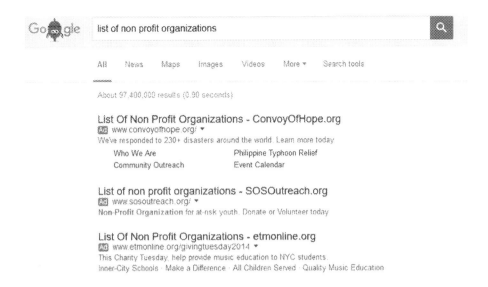

At the bottom of the page of Google search results is the link where you can go to the next page:

On each page, there is less competition because fewer people view the pages.

Now we return to the previous screen:

		Keyword	Status ?	Max. CPC ?	Clicks ?	Impr. ?	CTR ?
☐	✱	non-profits	☐ Eligible	$2.00	0	0	0.00%
☐	✱	not for profit organizations	☐ Below first page bid First page bid estimate: $2.69	$2.00	0	0	0.00%
☐	✱	list of non profit organizations	☐ Below first page bid First page bid estimate: $2.79	$2.00	0	0	0.00%
☐	✱	list non profit organizations	☐ Eligible	$2.00	0	0	0.00%

When a keyword is below the first page bid, that doesn't mean it won't display. It just means that based on the maximum bid that you've placed in AdWords, the ad is not likely to display on the first page of returned results.

▒ **Tip**   Google's keyword tool is nice to explore and it that can help you try out different scenarios for keywords. Check it out at `http://tinyurl.com/googkeywordtool`.

Conclusion

Dear Reader,

Congratulations on making it through the chapter!

If your head is swimming, the best advice I can give is not to worry about all the possible details and options. Just concentrate on trying the options that I've highlighted. Create an entire ad, breathe a sigh of relief, and launch back in. Remember to pay attention to your budget and set the start/end dates.

The next chapter takes a closer look at monitoring a campaign and checking performance.

Best wishes in your digital adventures!

Cheers,

—Todd

Monitoring a Campaign

This chapter looks at a live campaign, and I show you how I monitor it once you've set it in motion. Along the way, you'll learn more important concepts. The way I monitor my campaigns is not the only way to do it, but it works for me.

Setting the Report's Timeframe

Whenever I check on a campaign, the first thing I do is set the timeframe. When you get in the habit of doing this, you'll find timeframes that work for you, such as looking back at a week or a month. There's no right or wrong. Sometimes you're looking for trends to get a general sense of how things are going. In other situations, you may be reporting to a client. If you were reporting on a weekly basis, you could go in on Mondays and look at what happened during the previous week and then create a report for your client.

To change the timeframe of your report data, go into AdWords and click on the date range drop-down menu in the upper-right corner.

© Todd Kelsey 2017
T. Kelsey, *Introduction to Search Engine Marketing and AdWords*,
DOI 10.1007/978-1-4842-2848-7_4

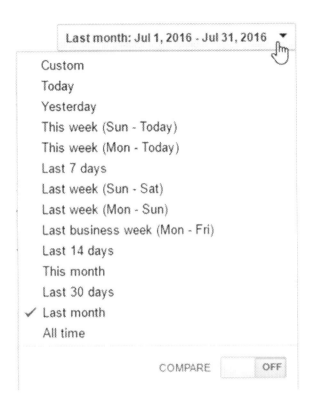

There are a number of default, common options, as well as a custom option at the top, which allows you set a specific range.

At the beginning of a month, I look at the last month to see how things went in the big picture. In this case, I'm looking at clicks.

Different timeframes enable you to look at trends and patterns. It can be helpful to look at a few months' worth of data, in order to see any patterns. Are there certain times of the week, for example, that seem to generate more (or fewer) clicks?

I also look at the previous seven days of data, especially when I'm reporting to a client on a weekly basis:

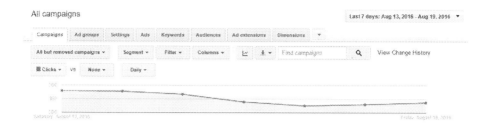

I commonly use the Custom feature, because I am not always checking AdWords the same day each week. In order to keep track of things, you can create a spreadsheet and track the last time you checked AdWords for a specific client. Then, the next time you check, you can look at the data from that date to the present.

For example, go to the drop-down menu and select Custom:

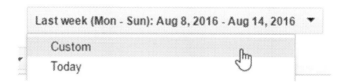

It displays a date range.

To set a custom date, you click on the first field (such as 8/8/16) to get the calendar to come up. Let's say that it's August 20, and the last time you checked the account was August 1. If you want to look at how the campaign went from August 1 through August 19, you click on the left date field and, on the little calendar, you select August 1:

The starting date is reflected in the drop-down menu. You can click on the date field on the right, the "ending" field (such as 8/14/16), to set the end date.

✓ Custom   8/1/16| -   8/14/16

In this example, we click on August 19. The tiny < and > arrows at the top of the calendar allow you to flip through the months if you need to.

If you're setting a custom date range, click the Apply button after you choose the end date.

Setting a timeframe is an important step, and you'll see it again and again. It's also an important mindset to have. One of the first questions you should ask yourself when looking at AdWords is, what timeframe am I looking at? Sometimes I forget to set the timeframe and at first glance, I'm confused by what I see. Then I realize that AdWords is just displaying the data I last looked at and I remember to set the timeframe.

If you haven't done so already, I invite you to adjust the timeframe in AdWords to look at various date ranges. Click on everything you can, change it, play with it, and just to get used to it. Explore AdWords! Remember to click on all those little question marks wherever you see them.

Tip Sometimes in order to *remember* to check AdWords, I sign in to Gmail (I highly recommend making it your primary e-mail address), then I go to Google Calendar (`calendar.google.com`) and create an event called "Check AdWords". I set the event to repeat and add an e-mail reminder. That way, on a weekly basis, I get a reminder to check AdWords. Maybe you're good at remembering, but these reminders are really helpful for me!

Clicks and CTR

Google—it's money for nothing and your clicks for free.

Well, I wish it was that way, but it isn't, so here we are. Someone's paying for the clicks. This is another reason to make sure you're checking not only how many clicks you get, but also the CTR (click through rate).

Basically, it's a matter of how many people click on your ad. Let's say that your ad has 1,000 *impressions*. That means that in theory, 1,000 people saw it. Now the question is, how many people actually clicked on it? If 10 people clicked on it, that would be 10/1000 = 1%. That is the click through rate.

The CTR is an indicator of how well individual ads or keywords are performing. Generally speaking, when I'm looking at a campaign, I take a general view. That is, I look at the campaign level.

When I'm looking at campaigns, I look at the clicks, and then I switch to looking at the click through rate. If I were last looking at clicks, I would see the following graphic. Then I click on the Clicks drop-down menu:

Then I choose Performance, and then CTR:

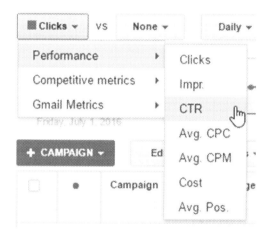

AdWords will show me the CTR—the overall click through rate.

I'm looking for how well it's doing. Is it okay? Is it going up? Is it going down?

This is a situation where I'll often select different timeframes. If I was looking at a month's worth of data, I might switch to look at two months of data, to get a larger picture.

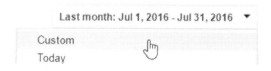

If I was looking at July 1-July 31, for example, I might switch and look at June 1 – July 31:

You can think of some of this as "zooming in and out" on the data, where you're looking at the information and forming an impression. I do more of this—look at different timeframes, different *metrics* (CTR, clicks)—when I'm trying to get a general sense of the campaign. When I'm reporting, I usually choose a single timeframe and metric to follow and often take a screenshot of clicks and CTR for the last week or month, for example.

While I'm looking at the data, in addition to the handy graphs at the top of the screen, I'm also looking at the *summaries*. In this case, you can see how many total clicks there were in that two-month period, and you can see the average CTR.

From that graphic, you can see that the General campaign is the main one—it gets most of the clicks, and it has a higher CTR. I put more effort into that campaign—it has more ads and the keywords have been optimized. The Global Recruit campaign is newer, with fewer ads and less optimization. It also targets a slightly different audience, and the keywords aren't pulling in as many clicks. The CTR is lower, but that's not necessarily a problem. Together, the average CTR across both campaigns is 2.11%. This is a respectable number.

On that note, there is no perfect CTR. Different industries have different benchmarks. Generally speaking, 1%-5% is good, but the more important aspect to work on is *improving* the CTR. That can be done through keyword research, as you'll learn in another chapter, or writing more ads to try different approaches.

For a learning experience, try searching for these phrases to see what you get:

- AdWords industry benchmarks for CTR
- AdWords what is a good CTR?
- AdWords how do I increase CTR?

Bookmark some of the links and read all you can. Don't be afraid to use Google often to learn more about how to do better at some feature or technique in AdWords. There are a lot of resources out there, for sure.

General to Particular: Campaign to Ad to Keyword

I review performance from the general to the particular, starting at the campaign level, then review the ads, and then look at the keywords. In most cases, when I'm reviewing a campaign, I start looking at the campaign level. It may the last view that AdWords remembers from when I last logged in; otherwise, I'll click on the Campaigns tab to get there.

When looking through campaigns, ad groups, ads, etc., it's easy to lose track of what "level" you're at. That happens with me; it might happen to you. To find your way again, go to the top of the screen to the Campaigns link, to get back to the highest level, and then "drill down" to the appropriate level.

After clicking on the Campaigns link, I click on the Campaigns tab if I'm not already there:

Then, I look at the individual campaigns. To look at a campaign, click on the campaign's name:

Google displays where you are at the top of the screen. At the Campaign level, you are shown any ad groups you have, and you can look at information just for particular ad groups if you like. Ad groups are a way to create ads and keywords that are closely related, to help you focus and ideally get more clicks, with a higher click through rate.

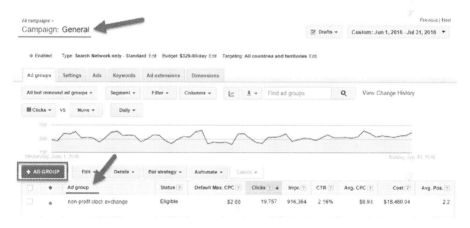

After looking at the high-level information about a campaign (and possibly choosing different time ranges to see the information on a campaign for different periods), the next thing I typically do is look at ads. You do this by clicking on the Ads tab.

You can see the performance of individual ads, such as the amount of clicks a particular ad is getting relative to others, or the CTR of one ad versus another. In the following view, there are only two ads. This is because I did try more ads, different variations of text, and so on, but then over time these particular ads seemed to generate the most clicks and have the highest CTR.

After looking at the ads, I switch over to look at performance of individual keywords by clicking on the Keywords tab:

After looking at the ads, I switch over to look at performance of individual keywords by clicking on the Keywords tab:

In some cases, this might be where the most "pruning" takes place, where you try a number of keywords and remove ones that aren't performing as well. I regularly take a look at keywords and re-rank the view, meaning I click at the top of a particular column and rank the results based on that item, such as CTR.

In this case, the ads are displaying with the lowest CTR first (.94%, for "stock investing"). This list of keywords also started out much longer, but with pruning, most keywords have already been removed that had less than 1% CTR. But I want to re-rank the results, so I click on the CTR column:

The keywords are now ranked with the highest CTR first, which in this case is "live stock quotes" at a CTR of 7.55%.

Keyword	Ad group	Status ?	Max. CPC ?	Clicks ?	Impr. ?	CTR ?
Total - all campaign ?				19,757	916,364	2.16%
live stock quotes	non-profit stock exchange	☐ Eligible	$2.00	12	159	7.55%
stock market recent news	non-profit stock exchange	☐ Rarely shown due to low quality score	$2.00	9	170	5.29%
at stock price today	non-profit stock exchange	☐ Below first page bid First page bid estimate: $2.84	$2.00	13	282	4.61%
sayer market	non-profit stock exchange	☐ Eligible	$2.00	20	437	4.58%
today share market rate	non-profit stock exchange	☐ Eligible	$2.00	11	242	4.55%
money share market	non-profit stock exchange	☐ Rarely shown due to low quality score	$2.00	53	1,191	4.45%

You can also rank the keywords based on the number of clicks. (Remember, when you're exploring, to click on all the little question marks! Question mark? Click on it!)

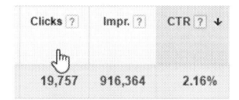

In this view, I see that even though it doesn't have the highest CTR, the keyword "stock market" has generated the highest number of clicks for the period of June through the end of July.

Keyword	Ad group	Status ?	Max. CPC ?	Clicks ?	Impr. ?	CTR ?
Total - all campaign ?				19,757	916,364	2.16%
stock market	non-profit stock exchange	☐ Below first page bid First page bid estimate: $2.38	$2.00	2,586	93,512	2.77%
stock market shares	non-profit stock exchange	☐ Eligible	$2.00	2,526	133,487	1.89%
stock shares	non-profit stock exchange	☐ Eligible	$2.00	2,125	104,667	2.03%
stock market stock	non-profit stock exchange	☐ Eligible	$2.00	1,936	79,549	2.43%
stock	non-profit stock exchange	☐ Below first page bid First page bid estimate: $3.14	$2.00	1,604	96,153	1.67%

Basically, you can think of reviewing a campaign's performance as taking a stroll through the grounds of the estate you've created, and if it makes sense, you could think of AdWords as a big garden, where you are trying different things—planting things and hoping they grow and produce results. Some areas will do better than others. The more attention you pay, the better things will do, and sometimes you need to try new things, different things. Sometimes you need to prune, trim, and remove things that aren't working.

AdWords is a garden!

Reporting

When you're managing campaigns, it's common to take a deeper look at various aspects of the campaign, and then depending on your client, colleagues, or manager, you report on a particular subset of what's going on. It is an educational process. When you explain more about how AdWords works to other people, they might be interested in more detail. But at a high level, reporting can be as simple as a single screenshot.

If you are reporting regularly, a typical option is to choose a weekly snapshot to report on how things are going. Reporting can also help you develop a strategy. You think of new things to try and then report on how things went. There's no right or wrong on how often to report, but reporting weekly and indicating a strategy for improvement on a monthly basis is a common approach.

I report based on how often clients want to hear from me. Generally speaking, most clients want to hear about how things are going on a monthly basis.

Sometimes I take a screenshot and insert it right in an e-mail; in most cases clients are fine with that. In other situations, reporting is more formal, such as placing screenshots or charts in an MS Word document along with commentary, and then perhaps including a few sentences as an executive summary.

To provide screenshots, I use Snagit, which I highly recommend. Other alternatives include the Snippet tool in Windows and the built-in screenshot capability in MacOS. One reason I like Snagit is because it's easy to edit screenshots and add text or arrows (I use it in this book).

I often end up setting the timeframe to a month and take a screenshot of the clicks for that month. Then I add a few relevant comments:

Then I include the numbers, or even a screenshot of the portion of the screen with the most relevant numbers:

Clicks ? ↓	Impr. ?	CTR ?
10,144	442,871	2.29%
480	32,306	1.49%
10,624	475,177	2.24%

Some clients like to know how the current time period compares to the last time period, such as this month compared to last month. You can set the timeframe for the previous month, get the numbers, and comment on how they went up or down. You can also set the timeframe for both months and show the differences in a screenshot.

To start out, I recommend keeping it simple. In general, I recommend playing a bit with each of the areas mentioned in this chapter. After you have a week's worth of data, try reporting on that week.

Conclusion

Dear Reader,

Congratulations on making it through the chapter!

You learned some of the ways you can look at information and monitor your campaigns. The next chapter takes a closer look at keywords, which are a critical part of creating a new campaign. They provide you with an opportunity for enhancing a campaign that's already in motion.

Best wishes in working with campaigns using AdWords!

Cheers,

—Todd

Keywords

This chapter explains how to research keywords so you can determine the ones that offer the best value for your ad campaigns. We also look at how you can adjust keywords, let Google pick additional related keywords, or limit that capability.

Getting Used to Creating Keywords

If you are intimidated by the idea of choosing keywords, or find terms like CTR and impressions confusing, I'm here to help. Being a numbers person can help, but you don't *have to* be a one to use AdWords. For example, look at me. You can see on LinkedIn that I'm a professor of business, and I teach digital marketing courses (http://linkedin.com/in/tekelsey—you're welcome to invite me to connect with you). You might think that I'm a numbers person, right? Not! I've learned what I need to know about numbers, but not so long ago I was a professional musician with a past life in rock 'n' roll (http://tinyurl.com/sistersoleil). I guess what I'm saying is, if I can do it, you can do it. After trying things out, you will get the hang of it.

When I started out using AdWords, I only tried maybe 10% of the available features. I mastered the basics to get going and then I gradually learned new features. In some cases, as you'll learn in this chapter, Google will offer you opportunities to learn about new features, right within the program. Google will analyze your campaign and come up with suggestions for things you haven't tried before, such as suggested keywords. Even if you never read another page of this book and only used the program (and clicked on all the question mark icons) and at least read and followed Google's suggestions, you will come a long way.

© Todd Kelsey 2017
T. Kelsey, *Introduction to Search Engine Marketing and AdWords*,
DOI 10.1007/978-1-4842-2848-7_5

Back to numbers—if you're a numbers person, great! I'm more of a words person. The reality is that if you've ever searched on Google, in some ways, you already know about keywords. That is, without even thinking about it, you've typed in things you're searching for, you've looked at Google search results, and chances are you've clicked on an ad from time to time, especially when you were looking for information about products or services.

At this point, take a deep breath. If you haven't do so already, take a minute to watch this excellent video by Matt Cutts of Google, titled "How Search Works." It helps give a sense of how things work on Google.

You can access it at https://www.youtube.com/watch?v=BNHR6IQJGZs or http://tinyurl.com/howsearchworks.

Another thing you might want to do to practice understanding how keywords relate to ads is to try a few more searches. Pick a web site or web page, anything really, go look at it, and then imagine what kind of keywords people would type in if they were trying to find that page.

For example, you could go on Amazon and search for "Todd Kelsey social media marketing" and find the page that has my *Introduction to Social Media Marketing* book.

Then, try imagining what you'd type in, or what you think people might type in, to find that specific book. Also, what kind of keywords would I want to use if the searchers didn't know about the book, but might be interested in buying it.

These are the two scenarios to about in relation to keywords:

- Situations in which people know about a product, brand name, or organization name

- Situations in which people are interested in a general type of product or service. You think of categories your web site fits into, which will help you think of keywords.

In the first example, you can type in something like "todd kelsey casa social media marketing" and you'll probably see something like this.

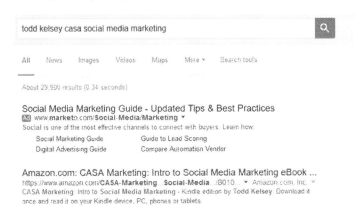

What I see is that there aren't any ads that are running for the book, at least on the first page of search results. But there is an ad running for a product called *Marketo*, which is a marketing automation platform. Nothing to do with me. What happened is that Google picked up the part of the search, "social media marketing," and the company Marketo is bidding on that search term. By the way:

Search term = Search query = Search phrase = Keyword

They all mean basically the same thing. Now back to Marketo. They sell an expensive marketing automation platform. At the moment, they happen to be casting their net wide on Google and using AdWords, with among many others, the keyword phrase "social media marketing". Even though it was only part of my search phrase, Google (and Marketo) thought I might be interested. It's something to remember.

If you look at the search results in the previous graphic, you'll see that the book does appear. However, the search result that is a direct match, in this case, is an organic search result. The top page is understandably on Amazon. So in this keyword example, a very specific search query like "todd kelsey casa social media marketing" leads to a pretty specific result. If I were running an ad, it would be okay to use that keyword, but not super critical, because the organic search result is on the first page of results anyway. Keep in mind that Marketo's pockets are far deeper than mine, and many companies bid on the phrase "social media marketing," so the price it would take to get an ad on the first page of results is probably not worth it to me.

Let's move on to the second scenario. I need to imagine keywords that would represent my book, whereby the searcher doesn't know about the book but might be interested in it if they did know. In that case, what topics or phrases would be appropriate to advertise my social media marketing book?

I rarely include written exercises in my books, but here we go, there's always a first time. Write down 3-5 phrases. These are phrases you could imagine someone typing in who might be interested in a book called *Introduction to Social Media Marketing* (Apress, 2017).

Okay, go!

_____

_____

_____

_____

_____

Need a hint? What topics does the book cover? Maybe these:

> social media marketing, learning social media marketing, intro to social media marketing

Try some of your keywords or use one of these hints and see what type of ads do come up.

Chances are your results will look something like this:

Master Social Media Marketing - Get Strategies, Tips & Metrics
www.marketo.com/Social_Media/Marketing ▾
Download our New 2016 Guide. Learn to create & execute a winning strategy!
Social integration tools · Nurture and score leads · Drive social ROI · Calendar management

| Social Marketing Guide | Compare Automation Vendor |
| Digital Advertising Guide | MA Buyer's Kit |

Expand Your Business Reach - With Social Media Marketing
marketing.dexmedia.com/social_media/marketing ▾
Learn How To Talk To Your Customers On Social Media With DexHub. Start Today!
Solutions: DexHub, DexLnk, Social Media, Websites, Search Engine Marketing (SEM)
Highlights: Over 3,000 Employees Dedicated To Your Success, Powerful Partnerships

1. [OPTIONAL] Introduction to Social Media Marketing
https://education.hootsuite.com/courses/**introduction-to-social-media-marketing** ▾
Course Level: Beginner. PLEASE NOTE: The Introduction to Social Media Marketing course is
OPTIONAL, and is not part of the Social Media Marketing ...

An Introduction to Social Media Marketing: A Guide for Small ...
www.**socialmedia**today.com/.../introduction-social-media-marketing-guide-small-busi... ▾
Jun 20, 2014 · It wasn't so long ago that social media was a completely new thing. Four years ago,
many people didn't know what social media was let alone ...

Social Media: The Free Beginner's Guide from Moz
https://moz.com/beginners-guide-to-**social-media** ▾ Moz ▾
The Free Beginner's Guide to Social Media from Moz has you covered. Learn best ... Chapter 1.
Introduction ... The sheer amount of data that customers make available through social media alone

There's that persistent Marketo again. There are a few big companies that an independent book author can probably not compete with based on the likely competition and bidding for those clicks. But the point is, if I wanted to, I could—it's open bidding season. There might be other, cheaper keywords that would work as well.

What's an AdWords learner supposed to do? Wouldn't it be interesting to see behind the scenes and know how much the clicks are going for? Remember, anytime someone clicks on a Google ad, someone is paying money. If it's an organic search result, it is just SEO (search engine optimization) at most and Google simply crawls the entire web classifying web pages.

SEM (search engine marketing) and SEO (search engine optimization) are closely tied together and both are important. You know they're tied together because they are both based on *keywords*. In one situation (AdWords), you pay for clicks. If your product is returned as an organic search result, you either got lucky, because the quality of your content was good, or you did some search engine optimization—or both.

Generally, as you think about keywords, your goal is mainly to get as high a rank as possible, that is, to appear on the first, second, or some such page of search results. (Of course, the more pages people have to click through to get to your search result or ad, the less likely it will get clicks.)

If you're willing to pay the price for the click, you can jump to the top of the list and have your ad appear on the first page of search results. There may be competition, which will drive the price higher, or you may get lucky and there is little competition.

The other aspect of search ads, or search results, to keep in mind, is search *volume*. There is a certain amount of demand for every keyword, from none to extreme. The more people type in a keyword, such as Olympics Rio, the more *volume* there is. Conversely, you might have a great keyword to represent your product, such as:

> joe's skateboard sticker kit rad gnarly neon dayglo
> starburst concrete warrior 10 pack

There might be very little competition for this phrase, and the price per click might be low, but is there much search volume? Is anyone actually searching for that very long phrase?

The point is, you can put all the pieces together, but you still need to have people searching. Like when you set your budget and set the maximum you are willing to pay per click—that doesn't mean you are necessarily going to use up your budget. The only way you use your budget is if people click on your ad. The only way that happens is if there's search volume. That is, people are out there searching for exactly the phrase you're bidding on, or something similar.

One final point—remember when you were creating ads? Did you take the time to look at the Google article about writing effective ads? You definitely should. The other aspect you need in order to get a response out of people, even if all the stars align—is good content. This is true even if you have your ad, keywords, cost per click, and everything nailed down. There are people out there searching for that phrase. People see the ad. But the ad still needs to be *compelling*.

Dealing with AdWords is like putting a recipe together. It's not any individual ingredient that makes the cookie—it's all the ingredients together, and they're all important. The quality of the ad is important, which includes how relevant it is to someone who is searching, how *well* it is written (which is why it's important to read Google resources and other resources on making effective ads), and how compelling the value proposition is. Is the item or service or web site unique in some way? Is there something you can brag about? Is there a discount or sale? All these things factor in. If the cookie looks good, people might eat it. If the ad looks good, people might click on it.

It's all about the clicks.

Wouldn't it be interesting to know behind the scenes about how much search volume there is for the keywords you're thinking about, how much competition there is for those keywords, and maybe how much the clicks cost? Well, lucky for you, AdWords has a feature you can use to figure out some of those things. It's called the Keyword Planner.

Welcome to AdWords Keyword Planner

To access the Keyword Planner, log in to AdWords, go to the Tools menu, and select Keyword Planner.

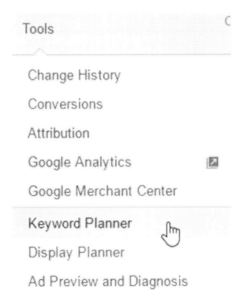

The Keyword Planner has several sections, and this section focuses on the first two, where you can find new keywords and get search volume data.

Be sure to check out "How to use Keyword Planner," which you can access in the same area, or via these links: https://support.google.com/adwords/answer/2999770 or http://tinyurl.com/keyplanner.

Getting Ideas

To get keyword ideas, click on Search for New Keywords Using a Phrase, Website or Category and review the options.

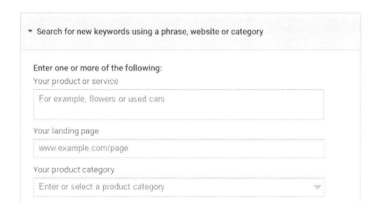

In general, you are entering some descriptive information—such as a type of product or service—to give Google some starting information. You can also input a web site or a specific page you are directing the ad to, so that Google can scan that specific page for keyword ideas.

Further down there are additional options, which you can skip for now.

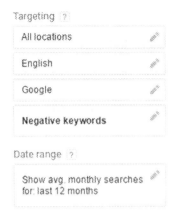 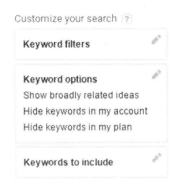

You could enter something like "digital marketing training":

When you're ready, just click the Get Ideas button:

Google will then give you some ideas, based on what you input. It will show information about the average monthly searches, including how much competition there is and what the bidding is going for. Remember to click on all the little question marks!

Don't be alarmed about the click price. There is an art and science to making AdWords "work" and hence this is where some of the competition comes in, especially when you are selling products. A company ends up deciding something like this:

1) We think we can get x number of people to click on an ad at a particular rate per click.

2) Of those people who click on the ad, we hope that some of them buy the product.

The simplified goal is to have a formula in which you are assuming that you need to spend money on advertising, and you expect to generate enough sales to cover the cost of advertising and make a profit.

Important Pause

Let's pause to consider this very carefully. This is the heart of Google, and the reason that Google makes $60 billion a year. Way back in the beginning of this book, we talked about return on investment (ROI) and about being able to track the impact and effectiveness of advertising. When you get down to these formulas, determine the cost per click, and track the impact, this is the magic of "trackable digital advertising". To review, it used to be you'd spend money somewhat wildly, advertising wherever you could, and hoped that people buy your products, but it was hard to know exactly how much revenue resulted from a particular advertising campaign.

But with Google ads, you can track *exactly* how much revenue you are making from an ad.

Google helps companies all over the world make billions and perhaps even trillions of dollars, with the confidence of knowing how effective their ads are. When you think of that formula, in the old style of advertising, you had to invest and then hope for the best. But when you are tracking ROI, you can invest and then know what works and what doesn't.

You might try a number of different keywords, ads, and approaches, and then find that you spend $1000 on an ad campaign, for example, and you know that $2000 of revenue results. That's valuable information to have.

Okay, now back to our regularly scheduled programming!

If you scroll down further on this screen, you get more ideas for keywords:

Keyword (by relevance)		Avg. monthly searches ?	Competition ?	Suggested bid ?	Ad impr. share ?	Add to plan
digital marketing course	⌇	22,200	High	$5.58	--	»
search engine marketing	⌇	12,100	Medium	$8.54	--	»
online marketing courses	⌇	4,400	High	$10.99	--	»
internet marketing course	⌇	1,600	High	$7.27	--	»

Google scans your site and listens to the information you're putting in. It comes up with suggestions. It also has an Add to Plan button (the >> button on the right), that acts like a shopping cart for keywords.

Just for perspective, many keywords fall into the .50 to $3.00 range, but the higher the price of the item, the more competition there is, and the higher the cost per click is. In the case of digital marketing, what is driving the cost per click up is that there are companies who want to sell higher priced services—such as software that costs several thousands of dollars a year, or courses that cost as much—so a book will have a hard time competing. It's just how the economics work.

Don't worry too much about the financial side. It's helpful to be able to talk about the power of digital marketing in terms of how it compares to other forms of advertising.

Search Volume

Search volume is another really interesting metric that you can look at. Remember the skateboard stickers? You might have a product or web site or keyword idea, and it would be helpful to know if people are actually *searching* for it. You can go into the Keyword Planner from the Tools menu:

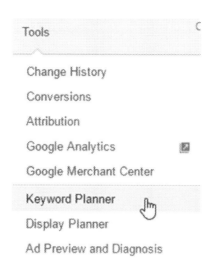

This time, click on Get Search Volume Data and Trends:

In this case, you can just enter your keywords and then see how many searches are there:

There are some additional options, but for now, I recommend ignoring them.

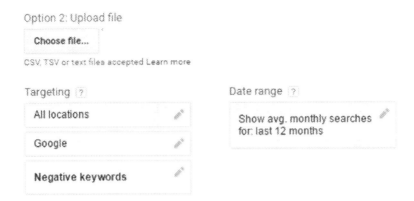

If you like, try thinking of a few keywords and then enter them as a test:

Then click Get Search Volume:

Google will show you the average number of searches, how much competition there is, and a suggested bid range.

Now, I'm going to ask you to do something on this area of the screen. Can you guess what it is?

You guessed it! Click on all the little question marks!

This is valuable information, and it can give you an idea of how to set your bidding, what your budget might be, and if you're selling products, to have information that could give Google a general idea of what the breakeven point might be.

In some cases you just have to try it and see what happens. But if you're particularly interested in the e-commerce side of AdWords, you'll want to look at *conversion tracking*, which tracks which ad clicks result in actual purchases. To track conversions, you enter a code that Google gives you. You can then track how much revenue results from particular campaigns, ads, or even from certain keywords.

When you have the conversion tracking in place, you can calculate how effective your ads are. For example, you spend x amount on the campaign, and it generated y amount of revenue. That's the power of conversion tracking, and when you get rolling with it and look back at keywords and bidding, you can generate a formula for an acceptable click price. You'll start to see of the 100 people who click on an ad, how many actually make a purchase. That's known as the *conversion rate*.

The bottom line is it helps to drive a lot of business, and it's an important issue for any company.

On a related note, if you are selling a service instead of a product, you might want to generate leads instead. Tracking the impact of advertising in this case is trickier, but it can be done. You might not track how much revenue is generated from an e-commerce purchase, but a business can still determine how valuable a lead is. It's a bit like a set of clicks when you're selling a product—of an x number of clicks, y of them result in a purchase. When you're talking about a service and about generating leads (people who might be interested), you'll have a "pipeline". You can then determine that if you get 100 leads, how many people will generally end up buying the service. You can then develop a similar formula, called the conversion rate, and be able to estimate how much each lead is worth. (whereby a salesperson might follow up with a lead). Based on how much the lead is worth, you get a sense of how much you can spend on getting clicks and new leads. That can help when you're looking at keywords and figuring out a budget.

Again, the best plan of action sometimes is simply to launch into it, try it, and look at the results. As with just about any topic in this book, there are a lot of resources out there, and you can Google a lot of good articles. Search for phrases like "conversion tracking with AdWords" or "setting up a lead pipeline with AdWords" and so on. There are a lot of good books and webinars, as well.

I don't want to scare you off with financial talk, but in the nitty gritty of keywords and bidding and cost per clicks, it's helpful to understand how *important* it all is. AdWords is a really important tool, and congratulations, it's a really important skill. Welcome to the party!

Activity

To get your head around some of these concepts and specifically the tools we cover in this chapter, I invite you to consider that you're working on a campaign for a client, and coming up with ideas. Try making a "search analysis" with suggested keywords for a web site. Pick a web site or product and do some research on keywords.

Sometimes AdWords is an individual activity, in other cases it is a collaborative process. If you are doing this for a client, sometimes they just want you to do your thing and let them know the results. Other times it can help to make a proposal and show that you've done some thinking. In either case, I suggest doing a little *keyword analysis*. Come up with some ideas and use the Google Keyword Planner to get a sense of search volume, competition, and the cost per click. You can share this information in a table, in screenshots, or however you like. Do some research and then present your options as you researched them, in a professional format of your choosing.

Conclusion

Dear Reader,

Congratulations on making it through the chapter!

You've taken a valuable dive into the world of keywords and clicks, including looking at how much they cost and how keywords relate to people's searching habits. Is there a lot of demand for just about every kind of keyword? You bet! Are there still some keywords to be unearthed in which the cost per click is low, competition is low, and it's a good deal for you? Certainly! Doing keyword research is a helpful thing to know and practice.

In the next chapter, we take a closer look at some strategies for working with campaigns.

Best wishes with keywords!

Cheers,

—Todd

Campaign Strategies

This chapter walks through a few helpful campaign strategies, including how to use Google's built-in Help function and its automated suggestions that can come up with ideas for your campaigns.

Getting Help

It's really important to remember that Google support is only a click or a call away. It's a great way to get free education, free assistance, anything you need.

Google will also automatically scan your campaign, and after it's running for a while, you might see an alert where it identified some ideas for keywords. It's all optional, but reviewing their suggestions can be a good way to learn.

When you're signed in to AdWords, you might see something like this:

There might be a little preview message and the little bell icon (notifications) might have a red exclamation point next to it. We'll come back to that later.

© Todd Kelsey 2017
T. Kelsey, *Introduction to Search Engine Marketing and AdWords*,
DOI 10.1007/978-1-4842-2848-7_6

Use the gear icon directly to access help. You might need to write your customer ID down (mine doesn't display here because I removed it from the image):

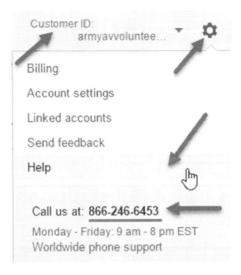

One particular point about this graphic I want to highlight is the toll-free number. (It might be different for you in your AdWords.) I definitely recommend putting it into your mobile phone so it is easy to call.

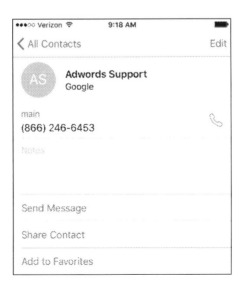

My experience has been that the hold times aren't very long, and you can always put your phone on speaker until someone answers.

If you click on the Help link, a window opens (try it):

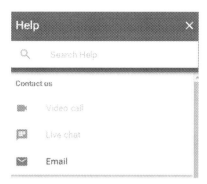

This is pretty cool. I've never actually tried a video call but that might be interesting. During business hours you can choose the live chat option. I often go with that because you get a more immediate answer than e-mail. But the e-mail option is nice too if it's not urgent and you have a general question about something (and you might even like to procrastinate about fixing it).

Support E-Mail

This section looks at a support e-mail. As an example, I asked a general question about a campaign, and I pulled an excerpt out of the response. I suggest reading through it once, and then we'll break it apart. The context is an ad campaign for www.armyav.org and the support person is recommending some specific things to try in this campaign, as well as sharing some basic campaign strategies. The point is you might want to try the support route to see what kind of suggestions they come up with, and as a learning experience.

> *I see that your ad in ad group is related to the history of the American military. It has always been recommended to have 4-5 ads in an ad group to see the performance of the ads. The ones that are not performing you can pause and then create new ads to replace them. Additionally, an ad group should have tightly themed keywords to increase relevance. As of now, I see that your keywords are related to your ads; however, current keyword match types may get you irrelevant clicks.*
>
> *The Keyword Matching option is a setting for each keyword that helps control how closely the keyword needs to match a person's search term in order to trigger your ad. As of now, you're using broad match keywords, which means the system allows your ad to show for searches on similar*

phrases and relevant variations, including synonyms, singular and plural forms, possible misspellings, stemmings (such as floor and flooring), related searches, and other relevant variations.

• Example: History army

• Searches that can match: History channel, army equipment, army games

However, using Phrase Match allows your ad to show only for searches that include the exact phrase, or close variations of that exact phrase, with additional words before or after.

• Example: "history of army"

• Searches that can match: history of army, our history of army, what is our history of army

Hence, I suggest you add more relevant keywords and consider using Phrase Match to improve the accuracy of your ads. Rest assured that you do not have to create different ad groups for different ads and keywords.

Okay, now I'll comment on this and maybe point out a few things:

I see that your ad in the ad group is related to the history of the American military. It has always been recommended to **have 4-5 ads in an ad group to see the performance of the ads**. The ones that are not performing you can pause and then create new ads to replace them. Additionally, an ad group should have tightly themed keywords to increase relevance. As of now, I see that your keywords are related to your ads; however, current keyword match types may get you irrelevant clicks.

They are highlighting the best practice of trying multiple ads. You create 4-5 ads in an ad group, let them run a week, for example, and then check back. You can pause those ads or delete them and make new ones. I call this pruning.

The Keyword Matching option is a setting for each keyword that helps control how closely the keyword needs to match a person's search term in order to trigger your ad. As of now, you're using **broad match keywords**, which means the system allows your ad to show for searches on similar phrases and relevant variations, including synonyms, singular and plural forms, possible misspellings, stemmings (such as floor and flooring), related searches, and other relevant variations.

Keyword match types are an important part of optimizing the performance of campaigns. It's a good thing to learn about. Starting out, I wouldn't be too worried about it—you can leave keywords to the default setting, which is Broad Match.

- *Example: History army*

- *Searches that can match: History channel, army equipment, army games*

*However, using **Phrase Match allows your ad to show only for searches that include the exact phrase**, or close variations of that exact phrase, with additional words before or after.*

It's worth reading the support explanation about Phrase Match and looking at the examples. The general principle is that to increase performance, you can narrow things down closer to what you want. Google will trigger an ad if people type in phrase *related* to your keyword, which casts a wide net, unless you tell Google otherwise. Broad Match casts the widest net, but some keywords that trigger your ad may be from somewhat unrelated searches. Optimization involves choosing more specific keywords and using the phrase match type.

- *Example: "history of army"*

- *Searches that can match: history of army, our history of army, what is our history of army*

Hence, I suggest you add more relevant keywords and can consider using Phrase Match to improve the accuracy of your ads. Rest assured that you do not have to create different ad groups for different ads and keywords.

The bottom line is that whatever question you have, there's a way to get an answer, whether through a phone call, a chat session or an e-mail.

Exploring Suggestions

I've found the suggestion feature very helpful, where Google scans campaigns and offers ideas and things you can try. Sometimes you see a message at the top:

You can click on the bell (notification) icon:

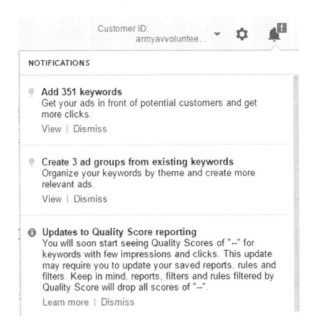

In this situation, you might click View in the Add Keywords section.

Google is basically suggesting a number of things you can do after scanning your site. Generally it takes a bit for a campaign to run before Google will suggest things. That's partly why it's good to check your campaign on a regular basis.

Google might come up with groups of keywords you can try.

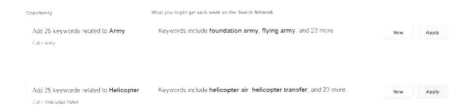

Then, you can click Apply if you want to dive in and see what happens. You can also dismiss the suggestion by clicking on the "x" to close the box.

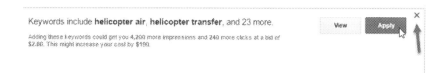

I recommend viewing the suggestion by clicking View:

There will be a list of keywords you can check or uncheck.

The context of this example is www.armyav.org.

Google will make suggestions for specific campaigns. If you accept the suggestion, Google automatically adds the keywords. Thanks Google!

Campaign: Call

Ad group: New | Existing

Army ▾

Bid: Default ad group | Keyword

$2.00

To accept the suggestion, click Apply:

Accessing Opportunities and Ad Groups

A variety of "opportunities" are available, sometimes from the beginning, sometimes over time as Google scans your campaign and web site. In some cases, they appear through the notification icon in the upper right.

You can also always click the Opportunities link at the top of AdWords.

Here's an example. I recommend clicking on the Learn More link, or any other link on the page:

You'll sometimes see suggestions, not just at a keyword level, but actually for particular ad groups. Remember the support person talking about ad groups?

The idea is to create a new container for several ads and choose keywords that are tightly related to the ads.

As with keywords, I suggest viewing first.

You end up with more learning opportunities—more Learn More links to help you understand what is being suggested until you're familiar with the process:

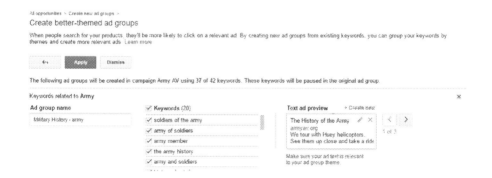

In this case, you can click Apply and follow through and try things out. Remember if you get stuck you can always call or chat with a Google rep.

DIY Ad Group Strategy

I think it's kind of fun and helpful to try the automated suggestions, but let's return to how you can adopt your own ad group strategy. To do it, create an ad group on a specific theme from something on a web site, make 4-5 ads, and choose a few closely related keywords.

To start with, I suggest creating an entirely new campaign. If you've lost your place in AdWords, just click on the Campaigns link at the top:

The context of this graphic is that clicking on the Campaign link places you in the Campaign view. After you create an ad group, you can access it under the Ad Group tab. If you're up for trying this out on your live AdWords account, remember that you can always pause your campaign after you've created it. You pause it by clicking on the little green button. If it's green, the campaign is active.

But you can click on it and go from Enabled to Paused:

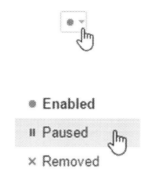

Enabled
Paused
Removed

Fighting the Wizard

Sometimes Google will put you through different wizards. I'm going to show you the "manual" way to create things, so when you create a new campaign, be aware that you don't have to follow the wizard:

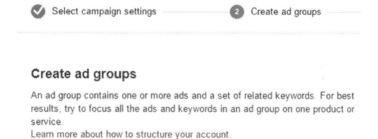

Create ad groups

An ad group contains one or more ads and a set of related keywords. For best results, try to focus all the ads and keywords in an ad group on one product or service.
Learn more about how to structure your account.

After you create your campaign, when it gets to the ad group level, you can click Save and Finish:

Continue to ads   Save and finish   Cancel

Then go to the Ad Groups tab:

Click the New Ad Group button:

Practically speaking, you might reach a point where you prefer to plan campaigns ahead of time. You can come up with ideas, use tools like the ones mentioned in the previous chapter, and develop with pages and links to promote. Then you can bring it in, all ready to go in AdWords.

Or you might prefer to try things "as you go".

When you get to the ad group level, if you're following this example, you'll want to think about a page on a web site to promote. It can be any site, because you can make an ad for anything you want. But the point is, take a look at your web site and pick a section or page to promote.

In this case, I'm looking at an earlier version of www.rgbexchange.org and have decided to highlight the Video section to create a "tightly targeted" ad group.

The section includes a free book available for download.

RGB: Investing in the World - How to Develop a Balanced Portfolio of Causes

This book shares the vision for the non-profit stock exchange, and talks about the idea of color coding causes. It can be viewed free in this section of the website, and is also available for purchase on Amazon in print and Kindle format.

iPhone/iPad/Tablets: If you'd like to read electronically, you can also view directly at this link: http://tinyurl.com/rgb-goog - this method allows you to scroll easily the the entire book.

The reason you want to look at a web site at this point is because when you create an ad group, you need to "point" it somewhere. The link you point your ads to is called the *landing page*.

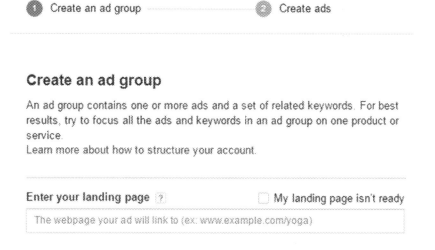

Create an ad group

An ad group contains one or more ads and a set of related keywords. For best results, try to focus all the ads and keywords in an ad group on one product or service.

Learn more about how to structure your account.

Enter your landing page ?   ☐ My landing page isn't ready

The webpage your ad will link to (ex: www.example.com/yoga)

In this case, the video page is found at the following link:

RGB: Investing in the World

That text at the top is what you can copy right out of your browser. It's the web address that you give to Google to indicate where the ad "points" to www.rgbexchange.org/book.

You can copy and paste a link from your web site (feel free to use www.rgbexchange.org as an example) and put it in as the landing page:

Create an ad group

An ad group contains one or more ads and a set of related keywords. For best results, try to focus all the ads and keywords in an ad group on one product or service. Learn more about how to structure your account.

Enter your landing page ?   ☐ My landing page isn't ready

www.rgbexchange.org

Google tries scanning the page, and it might come up with some suggestions that are relevant (or not):

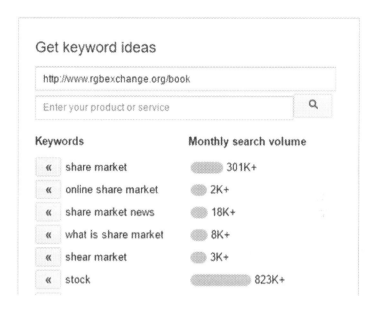

You can go to the bottom of the suggestions area and click on the **>** symbol to see more:

Hmmm, let's see—okay, that looks relevant:

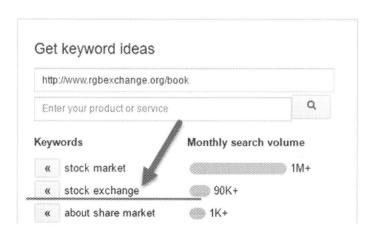

What you can do is click on the << symbol to "accept" the suggestion:

The keyword then appears in the Keywords section. (You can also type in your own ideas.)

Remember to click on the question marks! Always!

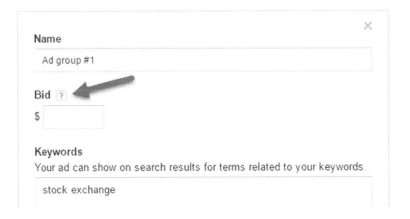

This is material worth reading:

> **The default bid is your maximum cost-per-click for ads in an ad group.**
>
> **How it works:** We'll use your default bid for clicks on your ads unless you set unique targeting-specific bids. The closer you get to your budget limit, the less your ads will show.
>
> **What to do:** You should set a bid that allows you to get the most value out of a click while staying within your budget.
>
> **Example:** On average, one in five clicks on your ad leads to a sale worth $5. Because you don't profit for click costs of $1 or more, you set a default bid of $1.
>
> More about choosing your bid and budget
> More about choosing a bid amount
> More about bidding by targeting method

There is no right or wrong about bidding. I suggest starting out with simple $1.00 bids. In my case I'm going with $2.00, because this AdWords account is a non-profit account and Google gives non-profits a free ad budget whereby the maximum bid is $2.00:

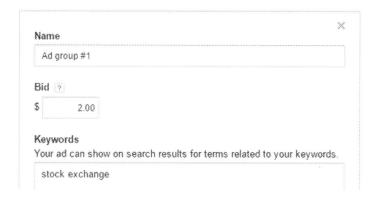

Now, when you're making an ad group and click in the keyword area to type in a keyword, there goes Google with more tips:

Tips
* Start with 10-20 keywords per ad group.
* Use match types to control which keyword variations can trigger your ad.
* Learn more about choosing effective keywords.

Try clicking in the keyword area, and by all means, read both articles. Remember the support person talking about match types? Here's another opportunity to learn about that concept:

About keyword matching options

Keyword match types help control which searches can trigger your ad. For example, you could use *broad match* to show your ad to a wide audience or you could use *exact match* to hone in on specific groups of customers.

This article explains the different match types that are available. Learn more about adding keywords.

You can access this article at https://support.google.com/adwords/answer/2497836?hl=en&authuser=0 or http://tinyurl.com/kmatchtypes.

Here also is the "Choosing Keywords" article, and lookie there, there's a video about basic tips for building a keyword list!

- You can view the article here: `https://support.google.com/adwords/answer/2453981?hl=en&authuser=0` or `http://tinyurl.com/choosekeywords`

- Here's the video: `https://www.youtube.com/watch?v=zFeOCCRpk8s&noredirect=1` or `http://tinyurl.com/5keywordtips`

Now scroll down a little further and you'll be ready to click Continue to Ads:

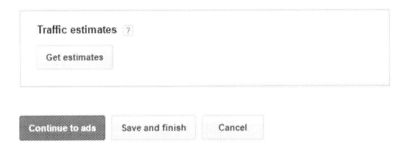

Don't be alarmed! Try it. If you want, browse through this chapter once, don't try anything, and then come back and try everything. Use the same web site as an example.

Remember you can always pause the campaign. You don't need to run a campaign—you can just create one and pause it. If you're following the example, enter something like this and check out how the preview forms on the right:

> ▒ **Note** Remember that Google changes things sometimes. This book is based on the traditional ad format, and if you go in and change ads in an existing account, they may look the same. But you may have access to a new format that allows you to enter more text, which is the one in the next diagram. Very similar.

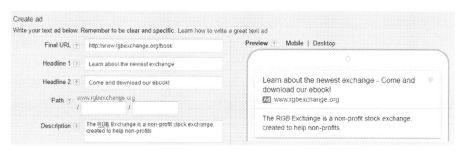

Click Create Ad when you're ready.

Woo-hoo! You've created your ad group. If you want, you can click the + sign and create more:

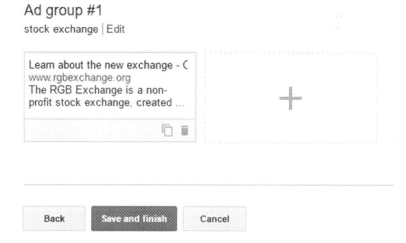

Then click Save and Finish.

Dealing with Errors

From time to time, you might encounter errors. Google might think one of your keywords doesn't fit its policies or you may have forgotten a detail:

Click Correct Errors and then look back at what you created. You will probably see some red text:

Doh! Okay, so I need to change the Ad Group name:

It's generally a good idea to name all your campaigns and ad groups to differentiate them.

Correct your errors if necessary and try clicking on Save and Finish again:

Remember, if you get stuck, help is only a click away:

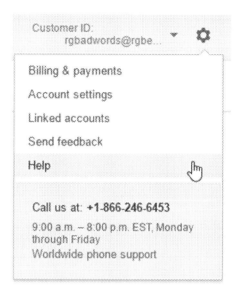

Finally, when your new ad group is created, you'll land back on the Ad Group tab.

▓ **Tip** Remember always to look at what date range you're using. Let the campaign run and then come and look back. If there's no activity, make sure you're looking at the right date range. (such as the last seven days or a custom date range of the time period since you started the campaign).

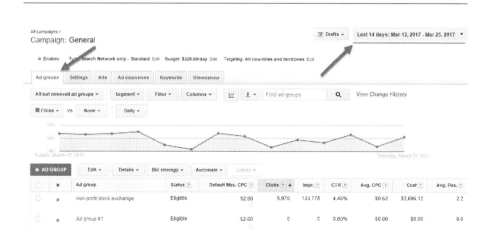

Conclusion

Dear Reader,

Okay. Whoo-whee! Congratulations on making it through the chapter!

This chapter took a whirlwind tour through building a few campaign strategies, asking Google for help, taking some suggestions, and building things out a bit further. In general, you can follow this approach and learn a lot, even if you never read another book! (Shhhh!)

The next chapter takes a closer look at a few ways you can improve performance through ad extensions.

Best wishes in your digital adventures!

Cheers,

—Todd

Ad Extensions

This chapter looks at a couple easy ways to make your ads more competitive. You can do this by adding extra links and information to the ads, through *ad extensions*. There are a variety of extensions you can use; we'll look at a couple that I've found helpful to get started: site links and callouts.

Search Results and Ad Extensions

To get a sense of what ad extensions can do for you, take a look at a sample search result (and try it yourself to see if you get something similar). Remember, the ads appear typically at the top and bottom of the page (red arrows at the top), and the "organic" search results (blue arrow at the bottom) appears *after* the ads. In this case, the first ad uses the *site link* ad extension to display links to sections of the site underneath the ad (underlined in blue).

© Todd Kelsey 2017
T. Kelsey, *Introduction to Search Engine Marketing and AdWords*,
DOI 10.1007/978-1-4842-2848-7_7

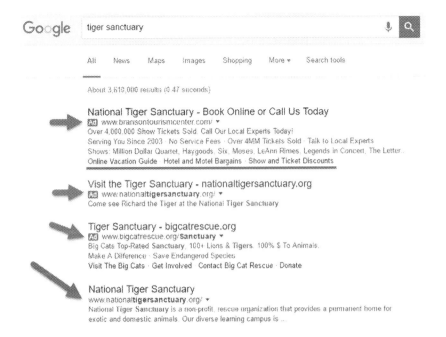

Site Links

Let's look at how to create a site link. Here's another example. The process involves deciding on the main sections of your site people that might be interested in:

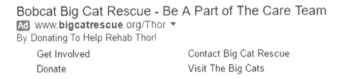

Keep in mind that one way to get information on site links is via the notification icon at the top; if you don't have any site links, it will sometimes suggest one and guide you through the process.

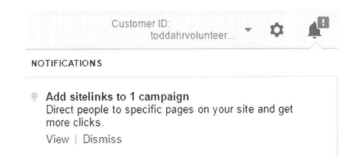

You can also do it yourself. In AdWords, go to the Ad Extensions tab:

▒ **Note**   If the Ad Extensions tab doesn't display for you, you might need to go to the right side of the tabs and click on the downward-pointing arrow tab. Then check the box next to ad extensions and click OK:

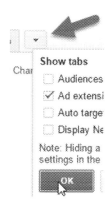

After getting to the Ad Extensions tab, you can click the selection drop-down menu and choose Sitelinks Extensions:

This shows a campaign that already has some site links. You can see how the site links perform. They tend to increase the number of clicks to your ads and raise the CTR (click through rate). In part because they're a way of getting people's attention.

Toward the bottom you see a section to add new site links, which also tells you which campaign has them.

To add a site link, click the + Extension button:

Then you can select a campaign to add it to:

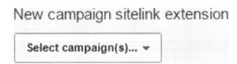

Click the little >> icon next to the desired campaign:

Click Done when you're ready:

Next, scroll to the bottom and click + New Sitelink:

➕ New sitelink

Site links fall into the category of something you should plan ahead for, along with ads, by identifying sections of the site that you want to highlight. Having a process or a checklist is also a good way to remember to add them. Do whatever you need to in order to remember to add them, because they can help your campaign and ads perform better.

In this case, try looking through the site and identifying pages you could refer people to:

About Us is a common one. Aim to get links and then think of 1-2 word phrases to represent them. Type in the phrase and then paste in the web address you want to send people to:

Then Save:

This is part one of the process. In the future, when you create new campaigns, you can go to the Ad Extensions ➤ Sitelinks screen and easily add saved site links to new campaigns, for a quick competitive boost to CTR.

In this case, since you're creating them, the default is to automatically add them. A list of available site links appears on the left, and since you just created it, the site link appears on the right. If it was on the left, you could click >> to add it. Likewise, if you want to remove a site link (on the right), you can click << to remove it:

You can create more site links if you want (by clicking on the + Sitelink button) or just finish up and click Save:

You will be returned to the General Extension screen. You can always get back to it by going to the Ad Extensions tab and then selecting the appropriate extension from the drop-down menu:

Part of your optimization process should be to try different text and links for site link extensions. This is similar to when you add more ads, look at their performance, and then remove the ones that aren't doing so well.

At the bottom the screen, you can also tell which campaign has which extensions enabled, and you can look at things on a campaign and ad group level.

I add extensions at the campaign level, but you can certainly do it at the ad group level as well (go to a campaign, choose ➤ Ad Group, and then choose Ad Extensions). Consider if you need a particular theme for an Ad Group, where you want to add site links that are especially suited to that campaign, to highlight the most related sections of the site.

Don't forget to keep your eyes open for suggestions from the Notification icon:

In this case, it will give you some information, and this example shows how the "suggestion wizard" tells you the value proposition:

It's suggesting that, by adding site links, you could get a .18% higher click through rate, which is a good thing:

What you might get each week on the Search Network

14 more clicks and a **0.18%** higher clickthrough rate.

The bottom line is that site link extensions are an easy win, and you should make creating them a habit or part of your checklist and planning process.

For More Reading

You may have noticed the Learn More link in the earlier graphic. Don't be afraid to click on these links!

Here is a link to the related help article: https://support.google.com/ adwords/answer/2375416?hl=en&utm_source=AWFE&utm_campaign=opptab or http://tinyurl.com/showsitelinks

They often have good, brief learning snippets.

Callouts

Another ad extension that's even easier to add and can help increase CTR are callouts. This is simply extra text beneath the ad (the fourth line in this case):

Acme Electronics
www.example.com
Shop ACME Electronics for laptops, smartphones, video games and more!
Free shipping · 24-7 customer service · Price matching

Often, it's a set of value propositions, such as "free shipping" and that kind of thing, but it can be any value proposition you want, including "low prices".

To try it, click on the Ad Extensions tab:

Ad extensions

Click on the selection drop-down menu and select Callout Extensions:

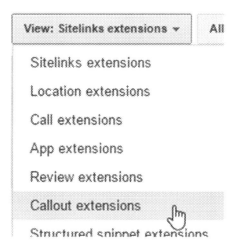

Toward the bottom, you'll see that you can add callouts at an account, campaign, and ad group level. To get started, click on the + Extension button:

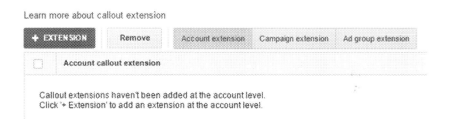

But wait—there's something else you can click on—that's right, another Learn More link. Do it!

Next, click the + New Callout button:

This is very easy. Again, as with site links, it's best to plan your callouts ahead of time, and it's almost certainly something you can do on a periodic basis. See what works and use that knowledge to try new things, especially when you add new pages or sections to your site (or even new add blog posts, for example).

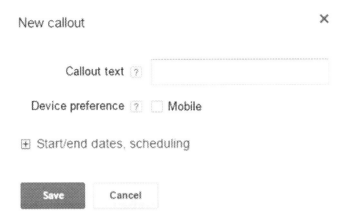

Start by clicking on the question marks to learn more. When you're ready, type in the callout text.

In my case it was a slogan. When you're done, click Save:

As with site links, when you're creating a callout the first time, AdWords puts them on the right, assuming you want to add them right then and there. If you don't, or you want to remove them, you can click the << button.

Take a few moments to add a few callouts. Remember that you can access them later through the Ad Extensions tab, and then by selecting Callout Extensions from the View drop-down menu on the left.

Conclusion

Dear Reader,

Congratulations on making it through the chapter!

In this chapter, we took a little breather and toured the world of ad extensions. After creating site links and callouts, you might want to explore the other ones, by clicking on the sections and options in the drop-down menu (under the Ad Extensions tab). Be sure to click on any Learn More links or question marks you see. You can go to the gear icon in AdWords in the upper right, select the Help section, and search for "ad extensions" or the name of the particular extension you want. Google help is pretty good, and if you want to learn more about the strategies related to the extensions, you might Google something like "adwords callout extension strategy" or "adwords site link extension strategy" and so on.

The next chapter takes a closer look at the wonderful world of getting certified in AdWords. It's a great thing—it's free, includes good learning material, and it's great to have on your resume or LinkedIn profile. If you're looking for work in digital marketing, certification definitely gets the attention of recruiters and potential employers. If you put the effort into it, it will make all your search engine marketing dreams come true!

Best wishes in your digital adventures!

Cheers,

—Todd

Getting Certified

Guess what? This chapter is really important. In this chapter, we take a look at how to get certified in AdWords.

AdWords Certification is a great thing:

- Free

- Good learning material

- Nice to have for advancing career

- Can help you get an interview

- Can help you get a job

- Doable!

Certification can be a nice thing to have on your resume or LinkedIn profile and you can brag to your colleagues. If you're looking for work in digital marketing, certification definitely gets the attention of recruiters and potential employers. If you put the effort into it, it will make all your search engine marketing dreams come true!

Why Get a Google AdWords Certification

I'll let Google step in for a moment.

> The Google AdWords Certification is a professional accreditation that Google offers to individuals who demonstrate proficiency in basic and advanced aspects of AdWords. An AdWords Certification allows individuals to demonstrate that Google recognizes them as an expert in online advertising.

© Todd Kelsey 2017
T. Kelsey, *Introduction to Search Engine Marketing and AdWords*,
DOI 10.1007/978-1-4842-2848-7_8

Imagine having it as a bullet item on your resume:

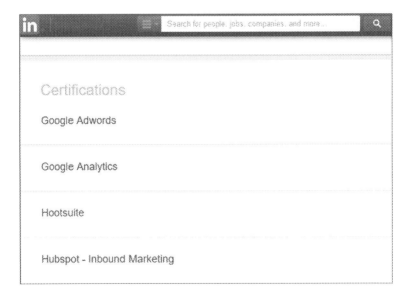

Imagine putting it in a prominent Certifications section on your LinkedIn profile:

There are lots of images available that you can use as well, whether on your web page, blog, company web site, etc.

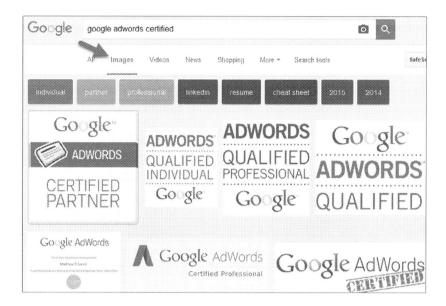

The bottom line is that it inspires confidence, and for good reason. It's not easy, per se, but it is doable. It shows you have some idea of what you're talking about. Probably my favorite part about it is that it is free. Many certifications cost money, but the AdWords Certification is free, well-known, and respected.

The Google AdWords Certification Process

To get certified, read through Google's free learning material on Google AdWords and then take the test. You can take it repeatedly (after an interval of a week).

To take it, you first need to register for Google Partners, at www.google.com/partners. Click on the Join Google Partners link at the top:

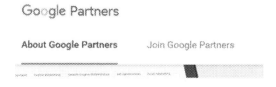

Then the Join Now button:

Let's be partners.

Join Google Partners and get access to free product exams and certifications, training events and promotional offers. Earn the badge to show your Google product expertise and specialization areas.

If you already have a Gmail address or a Google account, you can enter it and sign in.

Creating a Gmail Account

If you don't have a Google account, you can use an existing e-mail address such as blah@blipply.com. You can even use your work e-mail address to create a Google account. But if haven't tried Gmail yet, I strongly, strongly, recommend it, not only for easily getting in and out of AdWords and the certification, but also for all the other integrated tools, such as blogger.com and Google Drive/Google docs, which is like a free online version of Microsoft Office. Check out http://drive.google.com.

To create a Gmail account, just open a tab or a separate window and go to http://mail.google.com. Then come back and join/sign in to Google Partners with your Gmail address.

▓ **Tip** It' better to use a Gmail address, and if you want to still use another address, you can forward your Gmail to that address. (This is a good idea if you're just using the Gmail address for access to AdWords or the certification. Using e-mail forwarding helps you realize there's something you need to pay attention to. You can do it in settings in Gmail. Search for something like "Gmail how do I forward e-mails" in Google to get more information). You can also pull other e-mail into Gmail—such as from AOL, Yahoo, etc.

In other words, I strongly recommend making Gmail your central e-mail address.

Here are some reasons why this is smart:

- Comcast, Time Warner, Cox, BellSouth—these can change. If your Internet provider or work e-mail changes, you will still have your permanent Gmail address.

- Gmail's spam filtering is second to none.

- Google's Calendar is awesome and integrated.

- The ability to search old e-mails in Gmail is great.

- Google Drive has great tools like Google Docs, which is a free online equivalent to MS Office. You can compose and then download your files into other formats, or even invite others to collaborate on a document. It includes programs that mimic functions in Word, Excel, and PowerPoint. Super super helpful!

- Gmail is very mobile friendly.

Okay, okay! Enough about Gmail. I've just found it to be really helpful and I encourage all my students create a Gmail address.

Joining Google Partners

After joining and signing in, you need to accept the Terms of Service and then click the Next Step button:

Then, when you get to the sign up screen, as you're learning AdWords, I recommend choosing the Get Email option, and then clicking the Sign Up button:

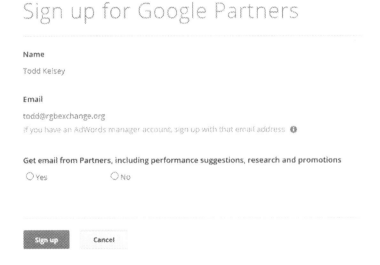

Next, you may want to take the tour:

Welcome to Google Partners ✕

Take this quick tour to learn about all you can do on Partners, including how to:

- Track your company's progress towards achieving Google Partner status

- Send promotional offers to clients and keep track of leads

- Get certified in AdWords

Click anywhere to start the tour

When you're signed in, the central screen in Google Partners allows you to access the certifications and the study materials. You can always sign in to google.com/partners and return to this same place. If you're following along, click the AdWords link:

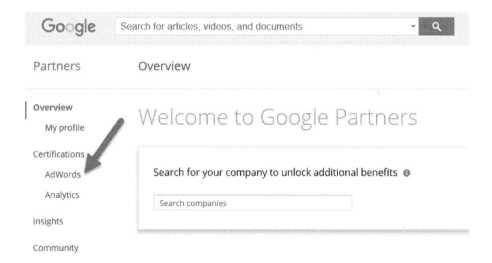

Here's what Google says:

> The AdWords exams cover basic and advanced advertising concepts, including campaign set up, management, and optimization. To become AdWords certified, you'll need to pass the AdWords Fundamentals exam and one of the other AdWords exams. You can demonstrate your expertise and help your company earn the Google Partner badge with an AdWords certification.

On the general AdWords Certification screen, the first thing you'll see is the AdWords Fundamentals exam. This is the first thing you need to take:

AdWords Fundamentals

This exam covers basic and intermediate concepts, including the benefits of online advertising and AdWords, and best practices for managing and optimizing AdWords campaigns.

Not yet attempted

Further down, there are other exams. If you want to go for it, do all of them. You only need to pass two to be officially certified. You have to pass the AdWords Fundamentals test, and I recommend the Search Advertising test as well, to make up the two exams you need for certification.

You need to pass two exams to get certified—the AdWords Fundamentals exam and one other exam.

The reason I recommend the Search Advertising exam is to keep your focus on the core of AdWords.

Search Advertising	Display Advertising	Video Advertising
This exam covers basic and advanced concepts, including best practices for creating, managing, measuring, and optimizing search ad campaigns across the Search Network.	This exam covers advanced concepts, including best practices for creating, managing and optimizing Display campaigns.	This exam covers basic and advanced concepts, including best practices for creating, managing and optimizing video advertising campaigns across YouTube and the web.
Not yet attempted	Not yet attempted	Not yet attempted

Shopping Advertising	Mobile Advertising
This exam covers basic and advanced concepts, including creating a Merchant Center account and product data feed and creating and managing Shopping campaigns.	This exam covers basic and advanced concepts of mobile advertising, including ad formats, bidding and targeting, and campaign measurement and optimization.

When you're on the main screen, you can roll your mouse over the AdWords Fundamentals section of the screen to read more details:

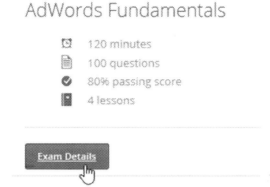

Click the Exam Details button. On the next screen, you can click the Take Exam button:

Further down, you can access the study material:

Remember to click on everything. This screen is the core section where you can access different modules. Notice that you can also download a PDF for more information. Consider having the PDF printed at Staples or some other office supply store, so that you have a physical copy to read through if your eyes glaze over.

Here is an example of the kind of articles you'll find in the Google Partners help section—see http://support.google.com/partners.

Remember, to be able to get back to any of this material, go to www.google.com/partners.

Planning Your Google AdWords Certification

Take the exam right now! I dare you! I actually recommend taking the certification exam just to experience it, partly to see what the questions are like. Start with AdWords Fundamentals and then, when you take the second exam of your choice, do the same thing.

Then, study for a few weeks (or a weekend, depending on how much time you can put in), take it again, see how your score is. If you go the extra mile and keep track of how much time you're spending studying and how much your score went up, you'll pass in no time.

Tips

I've taught a few AdWords classes, and generally from what I've seen, you get the most out of studying *when you are also using the program*. This means going through the study material from Google, but then also going into AdWords to look at and use the features.

Keeping It Real

It also helps if you are working on an active ad campaign. You don't need to have one that generates information, but it helps. You might set the budget to $5 a day for a week, and try to spend time on it each day.

Another strategy is to create the campaign and work through the features while it's paused. Technically you don't have to spend money to learn AdWords, but I strongly recommend creating at least one campaign where you do spend some money. Think of it as an investment in your career. You can actually try things like optimizing pages, taking suggestions from Google in the notification section, witnessing the boost in CTR you get from callouts or site links, etc.

Blogging About It

In the classes I teach about AdWords, I start by having people create a blog using www.blogger.com. Each week, students post 2-3 paragraphs, including a screenshot, about something they are learning. This technique can be a helpful way to capture and learn the information, and a side benefit is that you can also include your blog on your resume or LinkedIn page, or share it with colleagues, friends, and family.

Timeframe for Getting Certified

There's no universal answer about how long it takes—and it depends on what exam you are talking about. The Fundamentals exam takes the least time, and the rest are more intense. Pace you; it's doable. In the structure I teach, I typically have people work through one-two modules per week, and I think it's realistic if you're working. But if you can go through one or two modules a day, it's certainly possible. I've had people study hard and do it in a weekend—it just depends on your other commitments.

If you fall somewhere in the middle of the spectrum, especially if you're working and it's not something you can study for at work, I recommend picking an evening, or maybe early Saturday or Sunday afternoons, and going through a module. Write a small post about what you're up to on Facebook, or create a blog post and share that on Facebook, or message your friends on Snapchat, or whatever. Part of the point is that if you can share your progress, you'll find some folks cheering you on. The ideal situation is to find a study partner— someone else who wants to get certified.

Testimonials

> *"Last Friday at CareerFest, there were some digital marketing internships, as well as marketing internships where digital marketing skills were a plus. There was one internship for social media marketing. One company I talked to was really impressed that I was going for a certification in Google AdWords."*

> *"Professor Kelsey, I just wanted to let you know that I had a phone interview and the recruiter was really impressed by the fact that I am certified."*

If you're not in the job market—no problem. AdWords Certification can be just as valuable for advancement. Even if you're not sure you would work with AdWords at that company, it might be a good thing to learn. It's a $50 billion part of the digital marketing world, after all.

Inspiration

If you are in the job market, you might want to search for these:

- Salaries: `http://www.onwardsearch.com/career-center/ppc-jobs-salary-guide/`

- Specific job search: `http://www.indeed.com/salary?q1=adwords&l1=chicago`

Free Money from Google

Okay, it's almost free. Keep in mind that when you sign up for AdWords, you might get an e-mail with an offer for free ad budget. You can try signing up, poking around for a week but not spending any money, and then see what happens. Sometimes Google will e-mail you an offer or one will appear in AdWords—you might get a code for $100 of advertising credit. That credit is often activated after you spend $25. But that certainly can go a long way toward helping you get experience! Keep your eyes peeled. This is another reason to use Google Gmail when you sign up for AdWords and Google Partners.

Conclusion

Dear Reader,

Congratulations on making it through the chapter!

In this final chapter, we looked at how valuable and doable the AdWords Certification is.

Thanks very much for reading this book, and best wishes in getting certified!

Cheers,

—Todd

Special Request   Thank you for reading this book. If you purchased this book online, please consider going on where you purchased it and leaving a review. Thanks!

I

Index

© Todd Kelsey 2017
T. Kelsey, *Introduction to Search Engine Marketing and AdWords*,
DOI 10.1007/978-1-4842-2848-7

Get the eBook for only $5!

Why limit yourself?

With most of our titles available in both PDF and ePUB format, you can access your content wherever and however you wish—on your PC, phone, tablet, or reader.

Since you've purchased this print book, we are happy to offer you the eBook for just $5.

To learn more, go to http://www.apress.com/companion or contact support@apress.com.

Apress®

Printed in the United States
By Bookmasters